Keys to Chemistry

Book 1

Graham Hill

The English Universities Press Ltd

ISBN 0 340 19072 8 Pupil's Book 1
0 340 18075 7 Teachers' book 1

The English Universities Press Ltd
St Paul's House, Warwick Lane, London EC4P 4AH

Designed and illustrated by Richard Lewis and Christine Malone
Phototypeset by Filmtype Services Limited, Scarborough, Yorkshire
Printed and Bound for the English Universities Press Ltd
by Butler and Tanner Ltd, Frome
and London

Preface

This book is about chemistry and chemists and the roles they play in science, industry and society. I hope you will find much to interest you as you begin your studies.

No doubt your teacher will ask you to read various sections, but I hope you will read it 'for fun' also. There are pictures or diagrams on almost every page. You will be interested to look at these, but it is important that you *read* the text which will tell you more about them.

You will benefit a great deal if you follow up the ideas and facts in the book by discussion amongst yourselves, by trying out some of the suggested experiments and by reading other books you have at home, at school or in your local library.

Try to answer as many as possible of the questions in the text of each chapter because they will help you to obtain a really firm knowledge and understanding of the topics covered. There is a great variety of study questions so you should find plenty to interest and entertain you.

At the end of each chapter you will find a summary of the more important facts and ideas.

The practical book which accompanies this textbook contains a series of experiments which cover most of the ideas dealt with in the following pages. You could try some of the experiments at home.

Graham Hill
Feb. 1974

To Clare, Alison and Nicola

Acknowledgments

We are grateful for permission to reproduce photographs in this book as follows:

Cover:
Rocket: NASA
Breaking up solidified sulphur: Shell
Liming soil: Fisons Ltd
Chimney: Stewart Keating
Weaving: Hayward, Mitchell and Pavey Ltd
'The Alchemist', by David Teniers: Royal Picture Gallery, Mauritshuis, The Hague
Ink chromatogram: (Quink Permanent black – n-butanol: ethanol: 0·88 ammonia solution 3:1:1): Stewart Keating
Oil refinery: Petrofina (UK) Ltd

Page 6 British Steel Corporation 7 Turner & Newall Ltd 8 Tate & Lyle Ltd 9 (left) Crown Copyright Reserved – Geological Museum (right) Stewart Keating – courtesy of Prof R E O Williams, St Mary's Hospital Medical School, London 10 Trans Antarctic Association 11 (Figure 1.1) Reckitt & Colman Ltd 14 Thompson Newspapers Ltd 16 (left) Guiness (centre) Richard Lewis (right) Rowntree-Mackintosh Ltd 17 (Figure 2.4) Gallenkamp 18 United Nations 22 John Walker & Sons Ltd 23 (top) Roan Consolidated Mines (centre) Stewart Keating (bottom) British Steel Corporation 24 Dr J W Seakins, Institute of Child Health, Great Ormond Street Hospital for Sick Children, London 30 Richard Lewis 32 (top) Stewart Keating (bottom) Imperial Chemical Industries Ltd and Pfizer Ltd 36 Stewart Keating 40 British Lego Ltd 42 The British Museum 45 Richard Lewis 47 Maurice Nimmo 48 (top) Hallam Ashley (centre) British Aircraft Corporation (bottom) Upper Clyde Shipbuilders 49 (top) Richard Lewis (centre) Stewart Keating (bottom) Stewart Keating 54 Richard Lewis 55 British Insulated Callendars Cables Ltd 60 (left) Boots Pure Drug Co Ltd (right) Richard Lewis 61 Acheson Colloids Ltd 64 Sport & General Press Agency Ltd 65 (top) Radio Times Hulton Picture Library (bottom) Gas Council 67 (top) Mobil Oil Co Ltd (bottom) Barnaby's Picture Library 68 (top) South of Scotland Electricity Board (bottom) Stewart Keating 69 Evening Standard 70 United Kingdom Atomic Energy Authority 71 (top) Royal Institute of Chemistry (bottom) Radio Times Hulton Picture Library 72 Barnaby's Picture Library 73 Barnaby's Picture Library 74 Stewart Keating 75 Colchester Express 76 British Steel Corporation 77 British Rail Scottish Region 82 United Press International (UK) Ltd 83 The British Oxygen Co Ltd 84 Keystone 86 (first three) Stewart Keating (far right) Richard Lewis 88 Stewart Keating 91 Rothamsted Experimental Station 92 Sterling Health Products 94 Unilever 98 Ellard Educational Visuals 102 Thermit Welding (GB) Ltd 103 Griffin & George Ltd (Figure 8.8) Mount Wilson and Palomar Observatories 106 (top) British Steel Corporation (bottom) Dorset Natural History and Archaeological Society

Contents

Preface	3
1 What is chemistry?	6
2 Separating mixtures	14
3 Action of heat on materials	30
4 Elements – the building blocks	40
5 Action of electricity on materials	54
6 Burning, breathing and rusting	64
7 Oxygen – the reactive part of air	82
8 Competition among the elements	98
Index	109

Iron ore

1 What is chemistry?

1.1 Introduction

A heap of iron ore is useless as it stands. It looks like a dry dusty piece of earth: you can't eat it, you can't wear it, you can't even grow things in it. But if you heat it strongly with lime, coke and air, you can turn it into iron; and from iron you can get steel.

This is what chemistry is all about: finding out what the Earth is made of and changing useless things (like iron ore) into useful things (like iron and steel). More than that, chemistry is about knowing *how* and *why* you can make these changes happen.

Chemists call these different stuffs **substances.** Iron is a substance: so too are iron ore, lime, coke and air.

Substances that occur in Nature (those found on the Earth, in the sea, or in the atmosphere, and those dug out of the Earth's crust) are called *naturally occurring* substances. Substances which do not occur on the Earth but have to be made by chemical processes are called *man-made* substances. Iron and steel are man-made. Neither iron nor steel occurs on the Earth, but they can be made by chemical processes from iron ore. Asbestos occurs naturally on the Earth, but plastics like polythene and Perspex are man-made. Cotton and wool are naturally occurring fibres, but Terylene, Nylon and Acrilan are man-made fibres.

Try to answer these questions

1 Silver is a naturally occurring metal: it can be dug out of the Earth's crust (mined) as the metal. Name one other metal that can be mined as the metal.

2 Iron is a man-made metal: it has to be made and refined from iron ore. Name two other metals that have to be made from their ores.

3 Which of the following substances are man-made and which occur naturally?

Aspirin	Bronze	Coal
Steel	Sand	Bakelite
PVC	Coke	Diamond
Natural (North Sea) gas		

1.2 The chemist's source of useful raw material

Your mother is limited in what she cooks by the food she has in her kitchen. In the same way, chemists are limited in what they can produce by the raw materials they can get from the earth.

Asbestos is naturally occurring

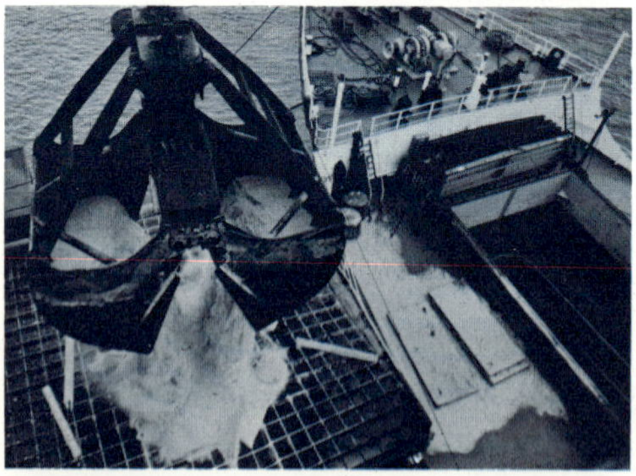

The chemist's most important raw materials are:

Plants
Coal and crude oil
Rocks and mineral ores
The sea
The air

PLANTS provide the bulk of our food and much of our clothing. Flour, vegetables and sugar all come from plants, as do linen and cotton. Rubber comes from the rubber tree, and coconut and olive trees give us oils for cooking, cosmetics and soaps. Some drugs (quinine and cocaine) come from plants.

COAL AND CRUDE OIL contain hundreds of chemicals used as raw materials for other products. For example, man-made fibres are made from crude oil chemicals, and coal tar yields the raw materials for many dyes.

ROCKS AND MINERAL ORES give us metals and glass. Some metals, like gold and silver, are naturally occurring whereas others must be made by chemists from raw materials. Iron is made from red-brown iron ore (haematite) and aluminium is made from bauxite.

THE SEA contains in every kilogramme about 35 g of dissolved matter, 27 g of which is sodium chloride (common salt) and 4 g is magnesium chloride. Large quantities of sodium chloride are used for preparing and preserving foods, for curing leather and for making sodium and chlorine. Magnesium chloride is used to make magnesium, a metal widely used in alloys (see Section 4.8).

AIR is a mixture of gases. Chemists have extracted neon, argon, oxygen, nitrogen and other gases from the air. Most of the nitrogen is used to make ammonia, nitric acid and fertilizers. Neon and argon are used in filling electric light bulbs. Oxygen is used in oxy-acetylene cutting and welding, and in the manufacture of steel.

All the time, chemists are trying to find out what the different raw materials in the Universe are made of and how to change them into new and more useful materials. The work of

Sugar cane is harvested in the West Indies and then crushed in a local factory, to yield a sugar-containing liquid. Dried raw sugar is shipped to the UK for further refining chemists has led to the use of many new materials, such as plastics, detergents, fertilizers and medicines.

New substances are sometimes first made, or discovered, by accident. Well perhaps not entirely by accident . . . usually they were discovered because the chemist was very careful to *observe* and to *note down* everything he saw. It is very unlikely that you will discover anything new just yet, but if you study chemistry as a career you almost certainly will. You must learn to look very carefully at everything that happens and not to miss a thing.

1.3 Describing different substances

Learn to describe the different substances you are looking at, in a way that other people can understand. In order to do this make a list of **properties** of the substances.

Chemists use the word 'property' in a special way. For instance, here is a list of some of the properties of wood:

1 It is brown in colour.
2 It is a solid substance.
3 It floats on water.
4 It can be made to burn.

'Native' silver in calcite (calcium carbonate)

Part of Sir Alexander Fleming's notebook for 1917

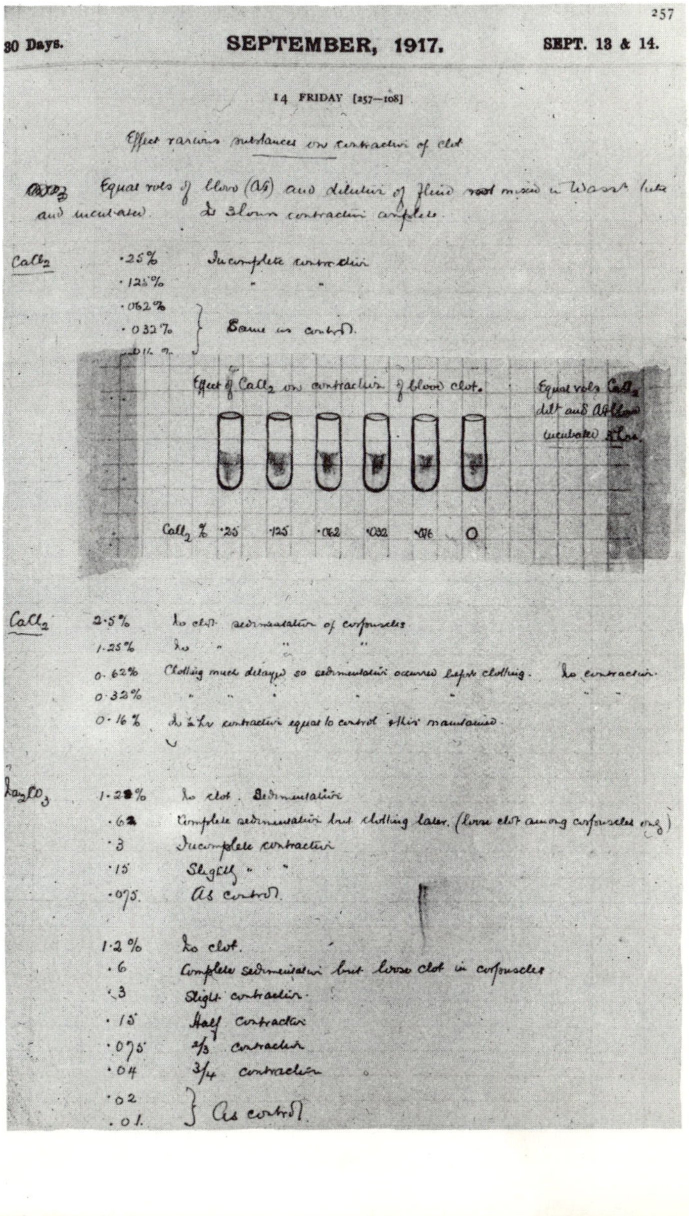

Properties of other substances can be listed in the same way. The four properties listed above are:

Colour

State There are three states in which substances exist, **solid, liquid** and **gas.** Sand is a solid, petrol is a liquid, air is a gas.

Density By this we mean how 'heavy' the substance is compared to the same volume of water. For example, wood is less dense than water.

Burning Learning about how things burn and why they burn is an important part of chemistry.

Try answering this question

List the properties of the following substances under the headings Colour, State, Density and Burning:

Iron Salt Ice Oil

If there are any answers which you do not know try to find out for yourself by experiment.

Can you think of any properties of these substances which we have left out? . . . **Smell.** As chemists, learn to use your nose as much as your eyes: but ve **very** careful.

There are two other properties which may be of interest:

1 Whether the substance dissolves in water.

2 Whether the substance can carry an electric current.

How many states of water can you see?

1.4 Solids, liquids and gases

Scientists find it easier to study different substances if they can sort them out in some way. They call this **classifying** them. At school you are sorted out, along with many other children, into different classes. All the children in your class are about the same age.

We must sort out the different substances we are going to study and put them in different classes. Substances with similar properties are put into the same class. By classifying substances in this way we can see patterns in the vast number of facts; this makes them easier to remember and easier to understand. One simple way to classify substances is to divide them into solids, liquids and gases. These are sometimes called the three **states of matter.** We call a substance which is a gas, such as air, gaseous. The gas burnt in your cooker is gaseous, too.

Sometimes a gas is called a **vapour.** For example when water evaporates we call it water vapour. The difference in meaning between the words 'vapour' and gas is very little. We use the word 'vapour' to describe a gas which can also exist as a liquid at the same temperature.

Try to answer this question

Classify the following substances as solid, liquid or gas:

Petrol	Chlorine	Salt
Chalk	Moth-balls	Steam
Oil	Iron	Vinegar

Add two more solids, two more liquids and two more gases to each list.

Can you think of a substance which is difficult to classify? Mud, for example, seems to be half way between a solid and a liquid. If you look at it more closely you will see that it is not a single substance at all. It is made up of sandy specks of a solid mixed with water. You may think smoke is a gas, but the soot which forms in a chimney is a solid powder. In fact, smoke is very tiny specks of solid floating in the air.

1.5 Pure substances and mixtures

Mud and smoke are not single substances, they are **mixtures** of substances. We have another way of classifying different substances: pure substances and mixtures.

The label in Figure 1.1 shows that the honey is pure. This means that it has had nothing added to it (or taken from it). Pure honey contains dozens of different substances; the relative proportions of these substances gives each honey its own characteristic flavour. But when the chemist says something is **pure,** he means that it is a single substance, and not a mixture of substances.

Try to answer this question

Classify the following as pure substances or mixtures. Put those you are doubtful about into a third class labelled 'don't know'.

Sea water	Soil	Brick
Air	Paint	Petrol
Soap	Milk	Disinfectant

Did you put any substances in the class labelled 'pure substances'? If so, I'm afraid you are wrong. You may think that soap and disinfectant are 'pure' since they are used for cleaning and killing germs, but to a chemist they are not single substances, but mixtures. In fact, all the substances in the list are mixtures. Petrol, for instance, looks like a pure enough liquid, yet it is a mixture of octane and other liquids. Air is a mixture of gases, mainly nitrogen and oxygen.

Figure 1.1

1.6 Separating pure substances from mixtures

How can we tell whether a substance is pure or whether it is a mixture? One method is to try to separate it into a number of different pure substances. The different 'ingredients' in a mixture are called **components.**

Sea water has two main components, salt and water. Air has two main components, nitrogen and oxygen. Petrol has several different components. When we talk about a component in a mixture we always mean a pure substance.

Summary

1 Chemistry is the study of substances around us in the Universe.

2 Substances which are found naturally are

called naturally occurring substances. Substances made by chemical processes are called man-made substances.

3 The most important sources of raw material are plants, coal and crude oil, rocks, the sea and the air.

4 The properties of a substance describe how it behaves under certain conditions.

5 Chemists classify substances by sorting out those substances with the same properties into the same class. Substances may be classified as solid, liquid or gas. Another method is to classify them as pure substances or mixtures.

6 The different substances in a mixture are called components.

Study questions

1 Have a look in your mother's kitchen and larder. Make a list of 10 of the substances and materials there. Use the chemical name for substances if you can. Which of these substances are naturally occurring and which are man-made?

2 One of the most important substances extracted by chemists, for use in medicine, is penicillin. Find out what you can about:
(a) the discovery by chance of penicillin by Sir Alexander Fleming.
(b) its extraction for medical use.
(c) the importance of penicillin.
(d) why it is becoming less effective in medicine as an antibiotic?

3 Which branch of chemistry are each of the following particularly interested in?

The biochemist	The pharmacist
The metallurgist	The geologist
The chemical engineer	

4 Chemistry has been described as a branch of science.
(a) How would you describe to someone of your age what science is?
(b) How would you describe to someone of your age what chemistry is?

5 Find out the names of the branches of science in which we study:

Living things	The stars and planets
The weather	Farming methods
Plants	Metals
Rocks and minerals in the Earth	

6 The chemical names of four substances we use at mealtimes are:

Sodium chloride	Acetic acid
Hydrogen oxide	Sucrose

What are their common names?

7 Coal is an important raw material. It is used as a fuel and many other materials can be made or extracted from it. Make a list of eight products made or extracted from coal.

8 Plants are a major source of raw material for the chemical industry. Make a list of six naturally occurring substances obtained from plants which are important in everyday life. Make a second list of six materials made from raw materials in plants.

9 Bloggs, a narrow minded young student, was asked by his teacher to explain what chemistry was about. 'The study of things out of bottles,' Bloggs replied. The teacher was shattered. Write 2 or 3 paragraphs to convince Bloggs that chemistry is concerned with everyday life.

10 (a) Find out which metals are obtained from the following ores. Give the chemical name for each ore:

Bauxite	Copper pyrites
Dolomite	Cinnabar
Haematite	Galena
Argentite	Calamine
Cassiterite (tinstone)	

(b) What do we mean by the word 'ore'?

11 All scientific discoveries can be used either for constructive or destructive purposes. For example, nitric acid is used in the manufacture of fertilizers and explosives. What other areas of chemistry present similar problems to modern chemists? Write two paragraphs on three of the following to show first their helpful and second their harmful uses:

Detergents	Atomic energy
Chemical warfare	Insecticides
The contraceptive 'pill'	

12 A French chemist and historian in the nineteenth century, C. A. Wurtz, Professor of Chemistry at the School of Medicine in Paris, began his book *The History of Chemistry* with the following words: 'Chemistry is a French science. It was founded by Lavoisier who will be remembered forever'.

(a) Find out what contribution Lavoisier made to chemistry in the eighteenth century.

(b) Write a letter to Professor Wurtz in reply to his claim. Give a list of six great chemists of various nationalities with a short description of their important achievements to show him that chemistry owes its success to men and women of all nations.

13 Have a good look at today's newspaper. What news or advertisements does it contain which are important to chemists and the chemical industry?

14 Write a short paragraph to explain each of the words below:

Substances	Property	State
Pure	Gaseous	Component
Classify	Mixture	

15 Make a list of some of the properties of:

Natural gas	Wool	Sand
Petrol	Chalk	Moth-balls

16 Write a few short sentences explaining to a friend who has not started to learn chemistry:

(a) why a chemist tries to classify substances.

(b) what a chemist means by a pure substance.

17 Apart from water, sodium chloride (salt) is the most abundant substance in the sea.

(a) How does it get into sea water?

(b) What is the simplest method of getting it from sea water in hot countries?

(c) Why is this method not used in this country?

(d) In this country most salt is obtained from inland beds beneath the surface of the land in Cheshire. How did the salt come to be there?

(e) More than 25 million tonnes of salt are mined each year in Britain. Many different substances can be obtained from salt. Most of the substances containing sodium or chlorine, which we use, are obtained from salt. Make a list of six substances which contain sodium or chlorine.

18 Set a candle burning. Describe carefully all you observe as it burns.

19 Have a look at some materials with a hand lens or under a microscope and describe carefully what you see. Try looking at:

A grain of sand	A grain of sugar
A human hair	A speck of dust

Crude oil

2 Separating mixtures

2.1 Introduction

Why do we separate petrol from thick black crude oil? Why do we separate pure water from sea water? Why are we interested in separating mixtures into pure substances?

Here are two important reasons:

1 We can get more, and more useful, materials. The pure substances can be put to uses for which the impure mixture was unsuitable. Just think what would happen if your father tried to run his car on crude oil rather than petrol!

2 Mixtures must be separated into pure substances before we can study the properties of the separated components and their principal uses. For example, only by doing experiments with pure water can we be sure that the reactions which occur are those of water and not some other substance it may contain. The properties of salty water are very different from those of pure water. Try drinking sea water.

The important question we ask in this chapter is 'How can we obtain pure substances from mixtures?'

Try to answer these questions

1 How can we obtain clear water from muddy water?

2 How can we obtain pure water from clear water which has solids mixed (dissolved) in it?

3 How are the various products (e.g., petrol, paraffin) separated from crude oil?

2.2 Decantation

Muddy water contains tiny particles of solid 'floating' in the water which makes it seem cloudy. If the water is left to stand for some time the heavier particles of solid sink to the bottom of the vessel and form a sediment. The clearer water can be separated easily from the solid sediment by pouring. This process, shown in Figure 2.1, is called **decantation.**

Decantation can be used to separate any liquid from heavier sediment. Have you watched your mother decanting water from vegetables after cooking?

Water obtained by decantation from its sediment of mud still appears cloudy due to very fine particles of solid 'floating' in it. We say that the particles are suspended in the water. 'Suspended' means 'hung up'. We call this mixture of fine solid particles suspended in water a **suspension.** In a suspension bridge, the road is 'hung up' with cables. Smoke is a suspension of fine solid particles in air. In a suspension there are two substances. The particles of the suspended substance are so small they do not sink even after a very long time. Have a look at some milk under a microscope. You will see tiny droplets of oil (cream) suspended in a watery liquid.

Figure 2.1

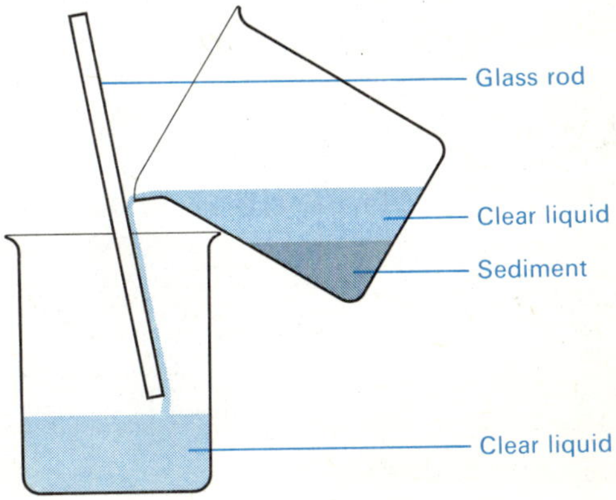

What are the components of these suspensions?

Try to answer these questions

Mist is a suspension.

1 What states of matter does it contain?

2 What are the suspended particles?

3 What are they suspended in?

How can we separate the suspension of minute solid particles from the cloudy water? How can we remove solid particles from smoke to obtain clean air? Think how your mother removes the water after boiling rice or peas. She strains off the water, trapping the solid rice or peas with a strainer. Can you think of other things used for straining off liquids?

Try putting muddy water through a strainer or sieve; you will not catch the solid particles because they are small enough to pass through the holes. You need a much finer sieve!

2.3 Filtration

We have a special word in chemistry which means 'to strain or sieve something'. We say that we **filter** it; and we call the process of filtering **filtration.**

In order to filter a suspension in which the particles are very small we must use something with even smaller holes to stop the solid particles from passing through. Filter paper, a special kind of thin white blotting paper usually made in the form of a circle, is made for this purpose. There are millions of fine holes in the paper through which water can flow, even though you cannot see them. Look at a piece of filter paper under a microscope – can you see the holes, or are they too small?

The simplest way of filtering is to use a glass funnel known as a filter funnel (Figure 2.2). The filter paper is folded so that it fits snugly in the funnel. The solid that remains in the filter paper is the **residue**, and the liquid that trickles through is the **filtrate**

Figure 2.2

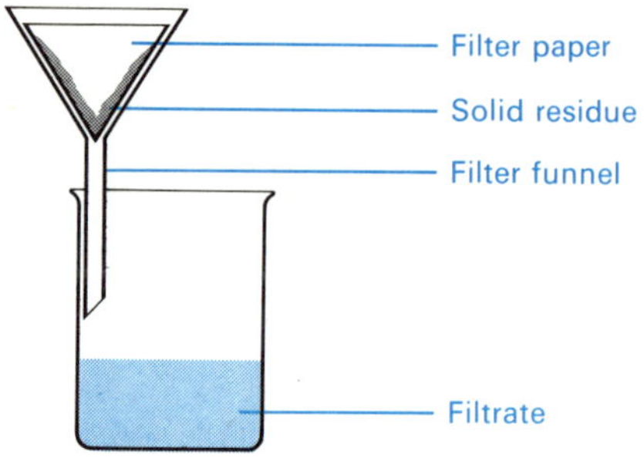

Figure 2.3 shows how river water is filtered before it is used for drinking and washing. First, the water goes through a sedimentation tank where the largest solid particles settle out; then it trickles through layers of gravel and sand in a filter bed.

Figure 2.3

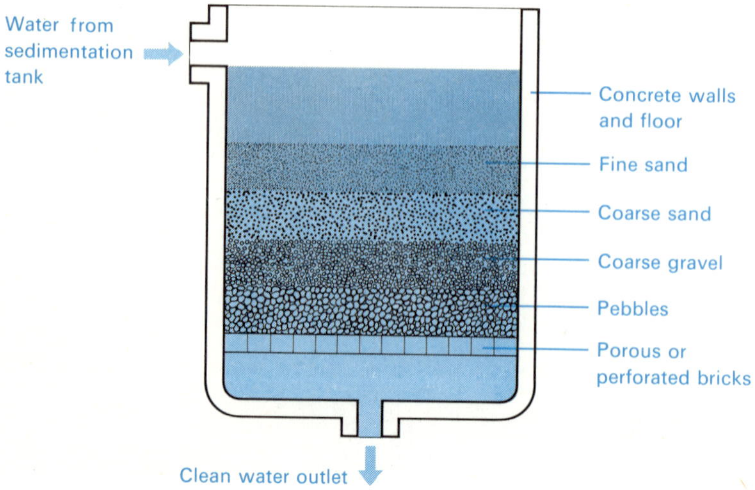

Any solid can be separated from a liquid by filtration. The liquid does not have to be water. 'Gloss' paint, for instance, is a suspension of coloured particles in a special liquid which contains no water. Try filtering some paint and see what happens!

Filtration is also used to separate beer from its sediment (yeast) before bottling. The beer is filtered by forcing it through filter cloths to catch the sediment.

Try to answer these questions

The household vacuum cleaner uses a process of filtration.

1 What forces the dust-laden air into the collection bag inside the cleaner?

2 Why can the air escape from the bag but the dust particles remain inside?

2.4 The centrifuge

The heavier solid particles in muddy water slowly sink to the bottom. Can we make them sink more quickly? The solid particles sink to the bottom because of the force of gravity acting on them. Can we make the force which pulls the particles down greater?

A machine called a **centrifuge** does this (Figure 2.4). The suspension is placed in a tube at the end of the arms; the arms whirl around very rapidly. This causes an enormous force which pulls the solid particles down to the bottom of the liquid. In the same way, a conker spun round on a piece of string pulls very hard on the string.

A spin dryer acts like a centrifuge and filter combined. As the dryer spins round, clothes and water are forced against the sides of the basket. The water passes through the holes in the basket which acts like a filter.

One of the most important uses of the centrifuge is in separating cream from milk. Milk is pumped into a bank of spinning, sloped discs. The heavier milk is forced down and outwards and the cream rises up and inwards (Figure 2.5). Sugar is being centrifuged in the photograph of refining on page 8.

2.5 Solutions and dissolving

What happens when table salt is stirred into water? What happens when sugar is stirred into tea? Does the salt or sugar form a suspension of solid particles?

The solid disappears provided you stir thoroughly and don't add too much salt or sugar. It is even possible to see through the salted water. This is very different from the

Figure 2.4

Figure 2.5

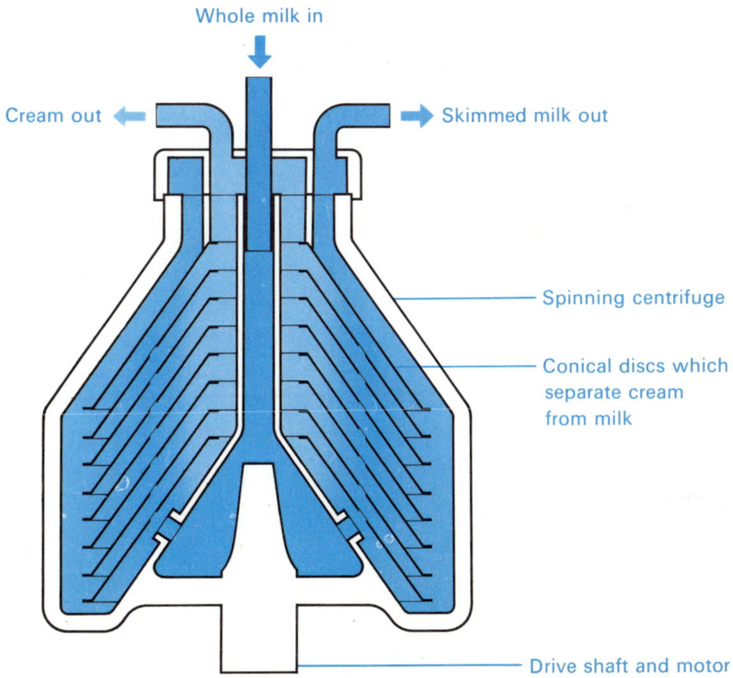

cloudy suspension which forms when mud is stirred with water.

How can we separate the salt from the water?

The simplest way to separate mud from water is by filtering or centrifuging. Try filtering the salt solution. The water which drips through still contains salt. (You can easily tell this by tasting it. But be warned! **It is usually very dangerous to taste any chemical.**) A centrifuge cannot get the solid particles of salt back again either.

The clear salt–water mixture does not seem to have any of the properties of a suspension. The solid can't be seen, but it must still be there because we can taste it. The solid has been broken up into tiny particles which are too small to be seen even with a microscope. These particles are so small that they can pass through the holes of the filter paper during filtration.

We say that the salt has **dissolved** in the water to form a **solution**. The substance which dissolves is called the **solute**. The liquid in which the solute dissolves is the **solvent**. Solids, such as salt, which dissolve are called **soluble**, and solids, such as mud, which do not dissolve are **insoluble**.

Try to answer these questions

1 Classify the following mixtures as suspensions or solutions:

Ink	Orangeade	Tea
Beer	Meringue	Steam
Exhaust fumes		

Harvesting salt from evaporated sea water

Figure 2.6

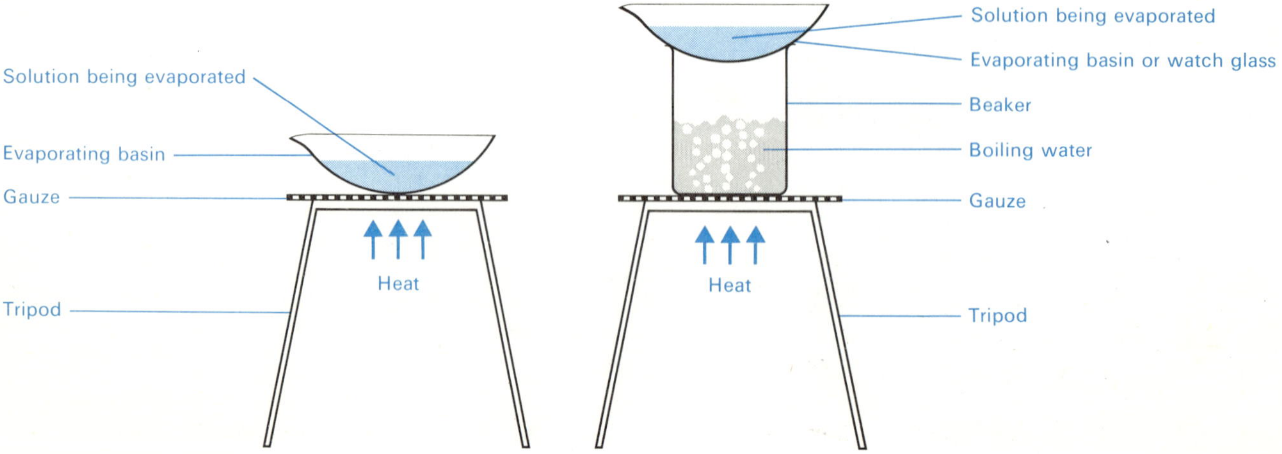

2 Are the following soluble or insoluble in water?

Sand	Sugar	Butter	Aspirin
Chalk	Oil	Vinegar	

Although water is a very useful solvent, there are many substances that are insoluble in water but soluble in other solvents. Oils and greases are insoluble in water but are very soluble in trichloroethane; this liquid is used by dry cleaners to remove grease stains from clothes. Nail varnish is insoluble in water and can't be removed by washing with water. Some brands of nail varnish remover contain a solvent called acetone, which does the job easily.

2.6 Evaporation and crystallization

We still haven't found a method of separating salt from salt water. In hot countries, sea water is left to dry; the water changes into water vapour and leaves behind the solid salt. Next time you are at the seaside look carefully for white rings of salt round edges of rocky pools.

This change of a liquid to a gas or vapour is called **evaporation.** Evaporation can be used to separate a dissolved solid from its solvent; raw cane sugar (page 8) is recovered in the crushing mill by evaporating off the water from the sugar-containing liquid.

The solvent evaporates and the solute is left behind. If the solute is left behind as solid crystals, we say it has **crystallized**.

Salt solution takes a long time to evaporate and form crystals. We can speed up the process by heating the solution gently over a bunsen burner or over a steam bath (Figure 2.6). If a solution is evaporated quickly the crystals left behind are usually small. Larger crystals can be obtained by evaporating the solution slowly.

Evaporation is an essential process in drying wet clothes and in producing concentrated evaporated milk. As the solvent evaporates from the solution the same amount of solid is left in less and less solvent. We say that a solution which contains only a little solute in a given amount of solvent is **dilute.** If a solution contains a lot of solute in a given amount of solvent we say it is **concentrated.**

As more and more solvent evaporates the solution becomes more and more concentrated. Eventually, the solution cannot become any more concentrated. If any more solvent evaporates some solid solute will come out of solution and form crystals. A solvent which has dissolved all the solute it can is **saturated.** As solvent evaporates from the saturated solution more and more solid crystallizes out.

Try to answer these questions

A saturated solution of sugar at 100°C contains 500 g of sugar in 100 g of water. At 15°C, a saturated solution contains 200 g of sugar in 100 g of water.

1 100 g of water saturated with sugar at 100°C cools slowly to 15°C, forming crystals of sugar. How much sugar does the saturated solution contain at 100°C? How much sugar does the saturated solution contain at 15°C? How much sugar crystallizes out of solution as it cools from 100°C to 15°C?

2 100 g of water saturated with sugar at 100°C is evaporated to dryness. How much sugar forms as crystals?

2.7 Distillation

We have seen how we can separate the dissolved solid from a solution by evaporating away the solvent. How can we collect the solvent which evaporates from the solution? For example, we can get salt from salt water, but can we also get pure water? When the salt solution is evaporated to a dry solid, the water escapes into the atmosphere as water vapour. If the water vapour is passed into a second container and cooled it will turn back to water. We have seen that all the solute stays in solution as the water evaporates, making the solution more concentrated. The water which evaporates and cools is thus pure water.

The process of cooling a vapour to a liquid is called **condensation.** The apparatus which causes cooling is a **condenser.** We call the process of evaporating a liquid and then condensing the vapour **distillation.** This means:

distillation = evaporation + condensation.

To speed this process up, we **boil** the solution. Boiling is the change of state from liquid to gas; the temperature at which a liquid boils is called the **boiling point.**

The apparatus in which pure water can be distilled from salt water is shown in Figure 2.7. Water vapour escapes from the boiling salt water and passes through the inner tube of the condenser which is cooled by the circulation of cold water around it. The water vapour condenses and drips from the lower end of the condenser into a receiver. The liquid which condenses and collects is called the **distillate.**

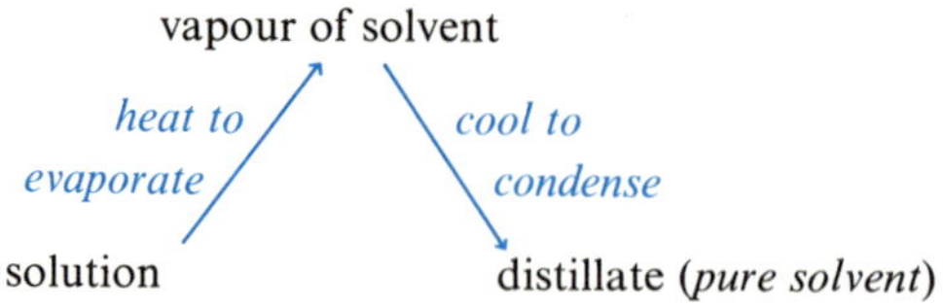

Distillation can be used to separate any solvent from its solution.

In most parts of the world tap water contains small quantities of dissolved solids. Pure water (distilled water) can be produced by distilling tap water or sea water. Most chemical laboratories have their own distillation apparatus (still) for producing distilled water. A school of 1000 pupils uses about 10 litres of distilled water each day. Distillation is also the essential process used in separating petrol, paraffin and the other components from crude oil. Crude oil is a complex mixture of liquids and dissolved solids. All the liquids in crude oil mix together; we say they are **miscible** liquids.

As the temperature of the crude oil rises, liquids with the lowest boiling point evaporate and form vapour most easily. Liquids that evaporate easily and have low boiling points are **volatile.** Thus, the first liquid to collect in the receiver will be the one that boils at the lowest temperature, i.e. the most volatile. Small amounts of any liquid which boils at a slightly higher temperature may also collect. The liquid remaining in the distillation flask will be richer in liquids which boil at higher temperatures. As the temperature rises further, liquids with higher boiling points will evaporate and collect in the distillate. Different portions of the original crude oil can be collected containing liquids whose boiling points are similar. The different portions are called **fractions.**

By repeating the distillation process with each fraction separately, it is possible to obtain a purer distillate. In this case, the process is called **fractional distillation.** At an oil refinery fractional distillation of the crude oil occurs continuously in the fractionating tower. Figure 2.8 shows a drawing of a fractionating tower, and the uses to which each fraction is put.

Mixtures of miscible liquids can be separated by distillation provided their boiling

Figure 2.7

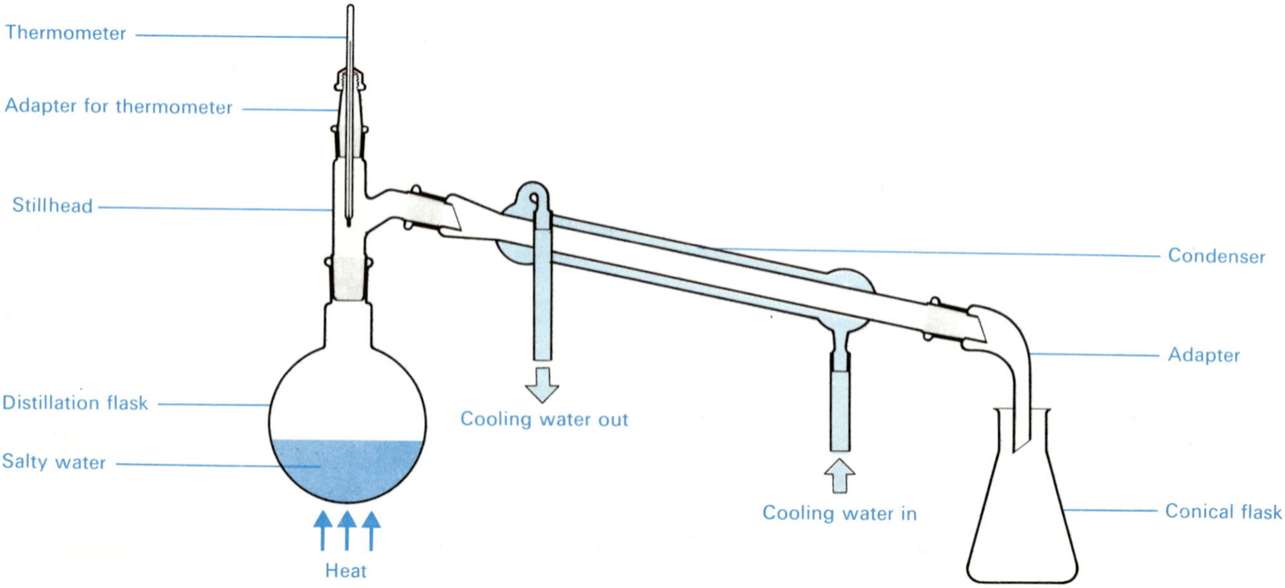

points are widely separated. Miscible liquids whose boiling points differ by only a few degrees can be separated by fractional distillation. Fractional distillation is also used in industry to separate oxygen from the other gases in air (page 84).

Try to answer these questions

Two other important processes which involve distillation are the manufacture of brandy and whisky. Brandy is made by distilling wine; whisky by distilling beer. The main components in both brandy and whisky are alcohol and water. The boiling point of alcohol is 78°C and that of water is 100°C.

1 Which liquid, alcohol or water, boils at the lower temperature?

2 Which liquid, alcohol or water, evaporates more easily?

3 Suppose some beer is distilled. Will the

Figure 2.8

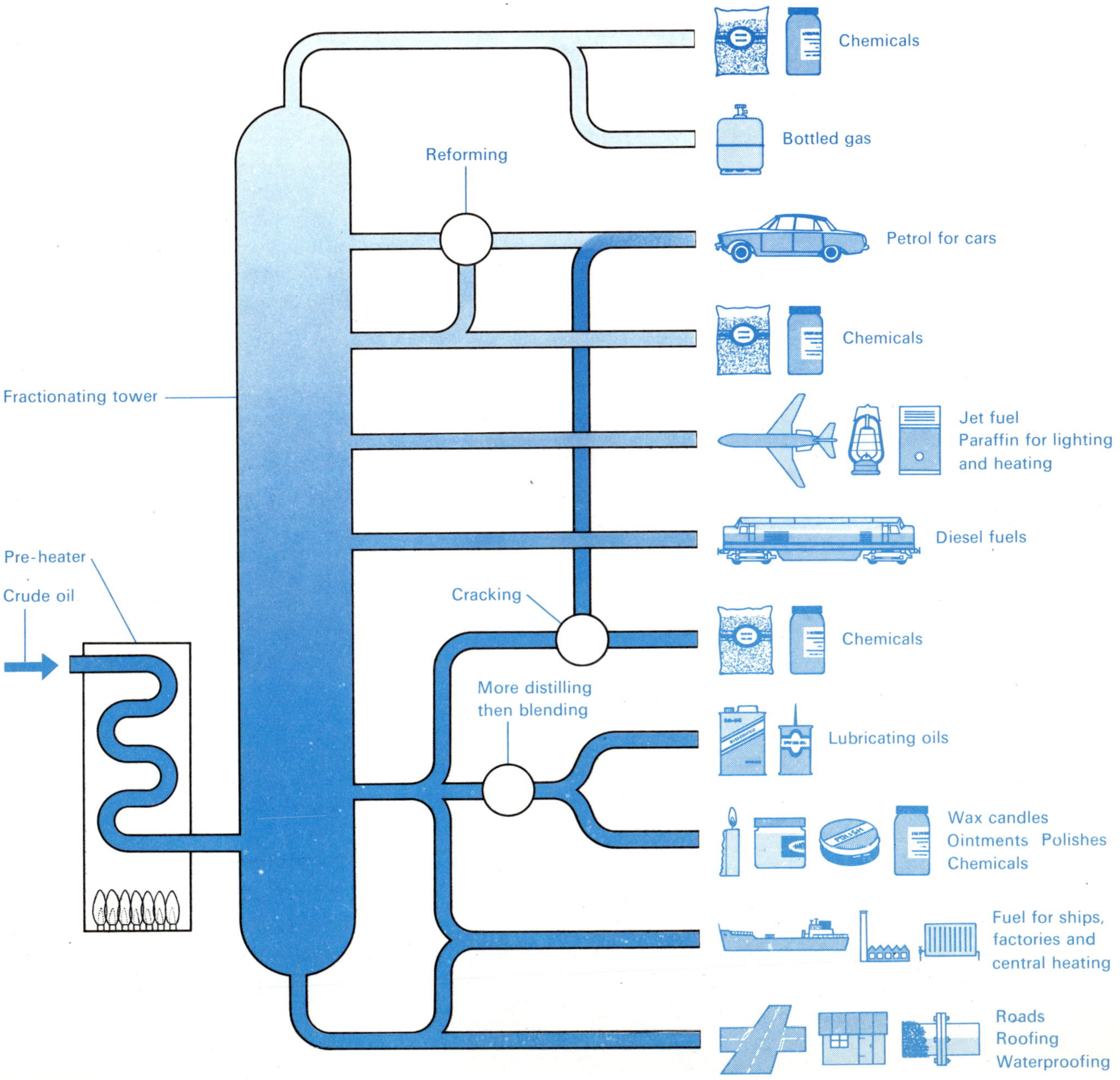

distillate contain a greater or smaller proportion of alcohol than the beer?

4 Does this agree with the fact that whisky is more alcoholic than beer?

2.8 Floatation

Liquids which mix together can be separated by distillation. How can we separate liquids which do not mix? These are called **immiscible** liquids. Cream and milk are examples of immiscible liquids; so are oil and water, molten fat and gravy. Each of these mixtures can be separated by decantation – pouring off the top layer. But, have you noticed how difficult it is to separate fat successfully from gravy or cream from milk.

Chemists use a separating funnel (Figure 2.9) to separate two immiscible liquids cleanly.

This method works because one substance is less dense than the other and floats to the surface. When the tap is opened, the more dense liquid runs into the beaker and is separated from the less dense liquid on top.

Try to answer these questions

(They illustrate how two insoluble solids can be separated by **floatation**.)

1 Is sawdust soluble in water?

2 Is sand soluble in water?

3 Does sawdust float or sink in water?

4 Does sand float or sink in water?

5 What happens when a mixture of sawdust and sand is thrown into water?

6 How can sawdust and sand be separated from a mixture of the two?

In this case, separation is possible if the two insoluble solids are added to a liquid in which one floats and the other sinks.

In industry, mineral deposits can be separated from denser earthy materials of clay and sand by floatation. The ores are crushed to a fine powder and then pumped to the bottom of a tank of water. Small air bubbles are blown through the mixture, which is also stirred. The air bubbles become attached to particles of crushed ore but not to the earthy material. The ore collects on the surface of the water as a froth, which is skimmed off.

Stillheads in a whisky distillery

Figure 2.9

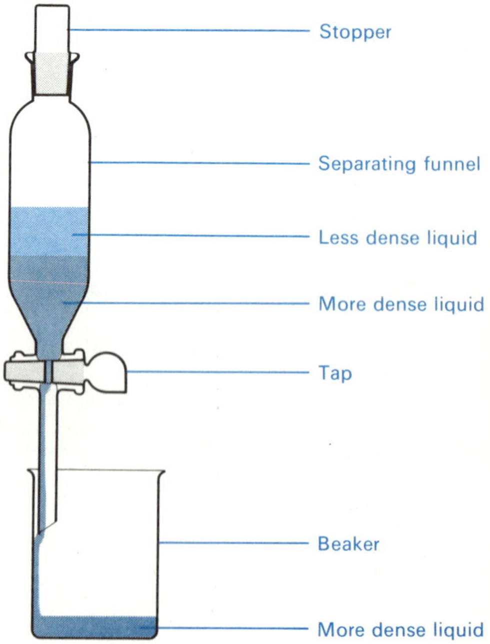

Copper-containing ore is skimmed off in the froth

Rocksalt

Electromagnetic separation of iron from other scrap metals. The large lump of lead in the foreground will be left behind

2.9 Key ideas in separations

If you have ever separated one thing from another, you will realize that they can only be separated if there is a difference between them. Cream can be separated from milk because of density difference. Mud can be separated from water because of the difference in the sizes of mud particles and water particles.

Separation of these mixtures is possible because of a difference in the properties of substances in the mixtures.

Try to answer these questions
Most salt in Britain is obtained from rock salt, a mixture of salt and sand, mined in Cheshire. How can the salt be separated from the sand? Think of a property in which salt and sand differ.

1 Is floatation on water (or any other liquid) possible? The density of sand is 2.6 g cm^{-3}, and that of salt is 2.2 g cm^{-3}.
2 Is salt soluble in water? Is sand soluble in water?
3 How can sand be separated from the solution of salt in water?
4 How can salt be separated from the salt water?

These questions show how to separate two solids when one dissolves in a solvent and the other is insoluble. Add the mixture of solids to the solvent, filter off the insoluble solid and then evaporate off the solvent from the filtrate to obtain the soluble solid.

2.10 Magnetic separations

How would you separate a mixture of similar sized brass screws and iron screws? Simple – just use a large magnet to pick out the iron screws. Large powerful electromagnets are used in industry to separate scrap iron from other pieces of scrap metal. In this case, the components in the mixture can be separated because they differ in magnetic properties.

2.11 Chromatography

If you have spilt ink on blotting paper you may have noticed that as the blot spreads out it forms a series of rings of different colours.

Figure 2.10

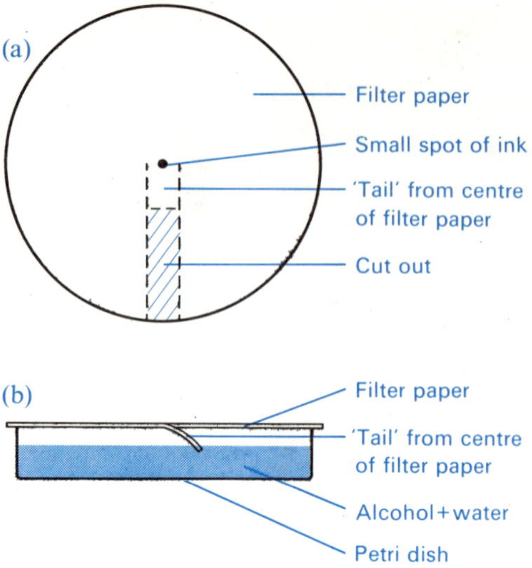

Each of the different coloured rings contains one of the dyes in the ink. Figure 2.10 shows how the individual dyes in ink can be separated more clearly using a filter paper.

Cut out the tongue of filter paper shown in Figure 2.10 (a) and fold back the 'tail' from the centre of the filter paper. Put a small spot of ink at the centre of the filter paper and leave it to dry. Now, place the filter paper on top of the Petri dish so that the 'tail' dips into a 50:50 alcohol : water mixture.

As the liquid soaks up the 'tail', the ink is washed through the filter paper, and the different dyes in the ink spread out in a series of differently coloured rings. Remove the filter paper before the liquid spreads to the edge of the filter paper, and leave it to dry.

Try the experiment at home using blotting paper instead of filter paper and water in place of the alcohol : water mixture. The different dyes can be seen better if you hold the paper up to the light. An even better solvent is n-butanol : ethanol : 0.88 ammonia solution (3 : 1 : 1), which will separate the components of black Quink in 15 min.

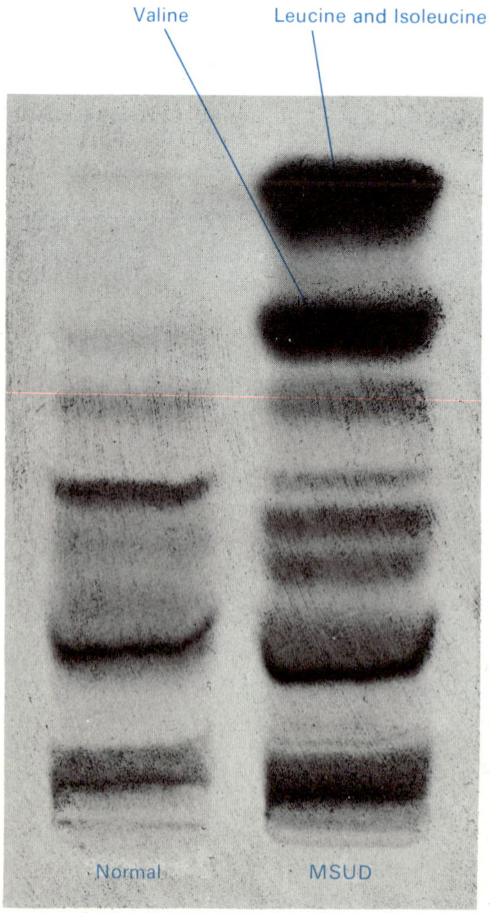

The photograph on the left shows a chromatogram of the urine of a normal child besides that of a child suffering from maple syrup urine disease (MSUD). The solvent was n-butanol : acetone : acetic acid : water (35 : 35 : 10 : 20 $^{v}/_{v}$). 4 or 5 babies born each year in the UK suffer from MSUD. Three of the amino acids produced when the body breaks down proteins accumulate in these babies. These three amino acids are leucine which is lethal, isoleucine which causes the babies to smell strongly of maple syrup and valine which is harmless. Chromatography of urine samples quickly reveals if these amino acids are accumulating. Treatment of MSUD involves eating controlled amounts of amino acids; without treatment babies suffering from MSUD will die within 1 month of birth.

Plants are another source of coloured substances. Grass is green and flowers are coloured because they contain coloured substances. The coloured substances in grass or flower petals cannot be removed by water, but can be removed by grinding the grass or petals with acetone or alcohol. Chromatography of the dyes extracted into the solvent can then be used to decide whether the coloured dyes are one substance or a mixture.

The first materials to be separated by this method were coloured and so the name **chromatography,** from the Greek word *khroma*, meaning colour, was used for the process. Nowadays the substances being separated need not be coloured. After chromatography has taken place, the paper is sprayed with a liquid which will react with the colourless substances to produce coloured substances.

Chromatography is very important in medicine; it can be used to separate drugs and other constituents in blood and urine. Blood samples from car drivers are examined by a special sort of chromatography; which shows not only if alcohol is present, but also how much there is. The natural colours of fruits, vegetables and meats can be enhanced by synthetic dyes. The use of these dyes is sometimes illegal. Chemists can detect these dyes.

What causes the dyes in ink and other substances to separate during chromatography? There are two properties in which the dyes differ and which cause them to separate:

1 Some of the dyes are more soluble in the alcohol:water solvent than others. Those dyes which are more soluble in the solvent move along the filter paper more quickly as the solvent moves along.

2 The dues are attracted to the paper, but not all dyes are attracted by the same amount. The effect of this attraction is to slow down the movement of a dye.

No two dyes are equally soluble in the solvent and equally attracted by the paper and so each dye moves through the paper at a different rate.

Summary of separation processes

Components of mixture to be separated	**Property in which components differ**	**Method of separation**	**Example**
Two solids, one soluble and one insoluble	Solubility	Add the mixture to the solvent. Filter off insoluble solid. Obtain the soluble solid by evaporation of the solution	Salt and sand
Two solids, both insoluble	Density	Floatation. Add the mixture to a liquid in which one solid sinks and the other floats	Sawdust and sand
A solid dissolved in a liquid	Boiling point	Obtain solid by evaporation to dryness or crystallization. Obtain liquid by distillation	A solution of salt in water
Two miscible liquids	Boiling point	Distillation or fractional distillation	Alcohol and water
Two immiscible liquids	Density	Separating funnel	Oil and water
Insoluble solid and liquid (possibly a suspension)	Size of particles	Filter or centrifuge off the solid	Mud or sand in water
Two solids; one magnetic, one non-magnetic	Magnetism	Use a magnet to remove the magnetic material	Scrap iron from other scrap metals
Solids or liquids	Solubility in solvent and attraction to the paper	Chromatography. Allow solvent to flow over the mixture, then extract substances from the paper	Separation of dyes

2.12 How can we test the purity of a substance?

If the separated substance is pure it will consist of only one kind of material. For example if the water we obtain by distilling sea water is pure it will contain only pure water. Its properties will be exactly like those of any other sample of pure water.

The temperatures at which a pure substance boils or **melts** are always the same. Melting is the change of state from solid to liquid; the temperature at which a solid melts (or a liquid freezes) is called the **melting point** or the **freezing point.**

The boiling point of a liquid changes with atmospheric pressure (water would boil at about 70°C on the top of Mount Everest). So we pick a standard pressure and measure all boiling (and melting) points at this standard pressure. Scientists have chosen the pressure exerted by a column of mercury 76 cm (or 760 mm) high as standard atmospheric pressure. This pressure is called 1 atmosphere (atm). In SI units 1 atm = 101.325 kN m^{-2}; 1 atm is also 1013.25 millibars (millibars are used to record atmospheric pressure in weather forecasting).

If our sample of distilled water is pure its boiling point should be 100°C and its freezing point 0°C at 1 atm pressure. The measurement of the melting point or boiling point of a substance is the best way to tell whether it is pure or not. In this way, we test the purity of a substance by comparing its properties with those of a pure sample of the same substance.

But, just suppose we don't know what our separated substance is. We cannot compare it with anything. How can we tell whether this unknown substance is pure? If the substance is pure, its properties will not change if we try to separate it a second time. For example, if we tried to separate pure water, we should have no success. The water obtained after separation would be exactly like the pure water we started with.

This idea gives us a method of testing the purity of unknown substances. If we have a pure substance its properties will not change after we have attempted to separate it.

Summary

1 The separation of pure substances from mixtures:
(a) enables us to examine the properties of single substances.
(b) provides us with many other useful materials.

2 A suspension consists of two substances. The particles of one substance are so small that they are suspended in the other substance and do not sink.

3 A solution is formed when one substance (the solute) dissolves in a second substance (the solvent).

4 Substances which dissolve in a particular solvent are soluble in that solvent. Substances which do not dissolve are insoluble.

5 Evaporation is the change of liquid to vapour. The process takes place at, and below, the boiling point of the liquid.

6 A dilute solution contains only a little solute in a given amount of solvent. A concentrated solution contains a lot of solute in a given amount of solvent. A saturated solution contains all the solute that can be dissolved in a given amount of solvent at the particular temperature.

7 Condensation is the cooling of a vapour to a liquid. Distillation is the process of evaporating a liquid and then condensing the vapour.

8 Liquids which mix together are called miscible. Those which do not mix freely are called immiscible.

9 The separation of a mixture depends on a difference in property of the substances to be separated.

10 The purity of a substance can be tested:
(a) by comparing its properties (usually melting point or boiling point) with those of a pure sample of the same substance.
(b) by showing that the properties do not change when a further attempt is made to separate it.

Study questions

1 The air used in a car petrol engine is filtered by drawing it through a pad of steel or glass wool coated with a thin layer of heavy oil.
(a) Why is it important to remove dust particles from the air before use in the petrol engine?
(b) What is the purpose of the layer of oil on the pad of steel or glass wool?

2 Some suspensions have particles that slowly float to the top of the liquid instead of sinking to the bottom. Milk is an example of this type of suspension.
(a) What happens if you spin such a suspension in a centrifuge?
(b) What are the suspended particles in milk?
(c) Find out what you can about the separation of cream from milk in the dairy industry.
(d) What products are obtained from cream after it has been separated from milk?

3 A centrifuge works better if it is balanced, that is, when loaded the arms opposite each other weigh the same as each other. A centrifuge is provided with containers for tubes on opposite sides of the central rotating shaft.
(a) What happens if the centrifuge is not balanced?
(b) Why does the centrifuge not work so well if it is unbalanced?
(c) Suppose you only have sufficient liquid to half fill one of the tubes of the centrifuge. What would you do to balance the machine?
(d) Does centrifuging have any advantages over filtration in separating an insoluble solid from a liquid?
(e) How is the liquid separated from the solid after centrifuging?

4 Explain clearly what you mean by the word 'suspension'.
(a) What is the usual state(s) of the suspended particles? What state(s) are these particles usually suspended in?
(b) Give three examples of suspensions from everyday life.
(c) What methods can be used to separate suspensions?

5 Draw a large, neat diagram of the apparatus you would use in the laboratory to obtain pure water from a sample of river water. Label the parts of the apparatus. How would you show that your product is pure water?

6 Your younger sister, playing 'shops' at home, decided to mix mother's sugar with some sand. Fortunately, you have a good chemistry set. Explain how you would separate dry sand and dry sugar crystals from the mixture.

7 Three solid substances have been mixed together. They are sand, salt and sugar. How would you separate dry samples of these substances from the mixture? Some properties of these substances are given in Figure 2.11.

8 (a) A sample of benzene has been obtained from coal. What experiments would you carry out to see if it is 100% pure?
(b) How would you find out if there is more than one coloured substance in blue ink?

9 What experiments would you carry out to show that concentrated orange squash:
(a) contains several dyes which make up the orange colour?
(b) contains water? How would you obtain a sample of pure water from the squash?

10 (a) True or false?
Water can be obtained from ink by:
(i) distillation.
(ii) evaporation.
(iii) chromatography.
(iv) filtration.
(b) Describe how you could obtain water from ink.

11 Evaporation is an essential process in drying wet clothes and in producing evaporated milk.
(a) What conditions (dry, damp, hot, cold, windy, calm) help evaporation of water

Figure 2.11

Substance	Density	Effect of water	Effect of alcohol
Sand (silicon oxide)	2.6 g cm^{-3}	Insoluble	Insoluble
Salt (sodium chloride)	2.2 g cm^{-3}	Soluble	Insoluble
Sugar (sucrose)	1.6 g cm^{-3}	Soluble	Soluble

from the clothes?
(b) What is evaporated from the milk?
(c) What is the difference between evaporated milk and ordinary milk?
(d) How do you think milk is evaporated industrially?
(e) Find out what you can about the industrial evaporation of milk.

12 A saturated solution of salt (sodium chloride) at 15°C contains 36 g of salt per 100 g of water. A saturated solution of salt at 100°C contains 40 g of salt per 100 g of water.
(a) What is the total weight of 100 g of water saturated with salt at 15°C?
(b) How much salt is required to saturate 1000 g of water at 15°C?
(c) What can you say about the amount of salt required to saturate 100 g of water at 30°C?
(d) How much solid salt can be obtained if 100 g of water saturated with salt at 15°C is evaporated to dryness?
(e) 100 g of water saturated with salt at 100°C is cooled to 15°C.
 (i) How much salt does the saturated solution contain at 100°C?
 (ii) How much salt does the saturated solution contain at 15°C?
 (iii) How much salt forms as crystals when the solution cools from 100°C to 15°C?

13 Copy out Figure 2.12 and fill in the gaps. (Chlorophyll is the green pigment in plants.)

14 A sample of crude oil was distilled in the apparatus shown in Figure 2.13. Four fractions of distillate were collected in the temperature ranges:
20°C – 70°C
70°C – 120°C
120°C – 170°C
170°C – 220°C
(a) The crude oil is not heated alone but with asbestos wool. Why is the asbestos wool used?
(b) Why is the receiving tube placed in a beaker of cold water?
(c) Why would it be better to use a condenser to condense the vapours?
(d) Which fraction will have been condensed least efficiently?
(e) Which fraction of the distillate will collect first?
(f) Which fraction of distillate could be used as petrol in a car?
(g) Which fraction of distillate is most flammable?
(h) Which fraction burns with the smokiest flame?
(i) Which fraction would be darkest in colour?
(j) Why is the asbestos black at the end of the experiment?

15 Figure 2.14 shows how a magnetic mineral ore can be separated from non-magnetic material. Explain carefully how the magnetic ore is separated from the non-magnetic material.

Figure 2.12

Solute	Solvent	Soluble
Salt	Water	Yes
Sand	Water	
Oil		Yes
Sugar		Yes
Chlorophyll	Water	
Chlorophyll		Yes
Naphthalene		Yes

Figure 2.13

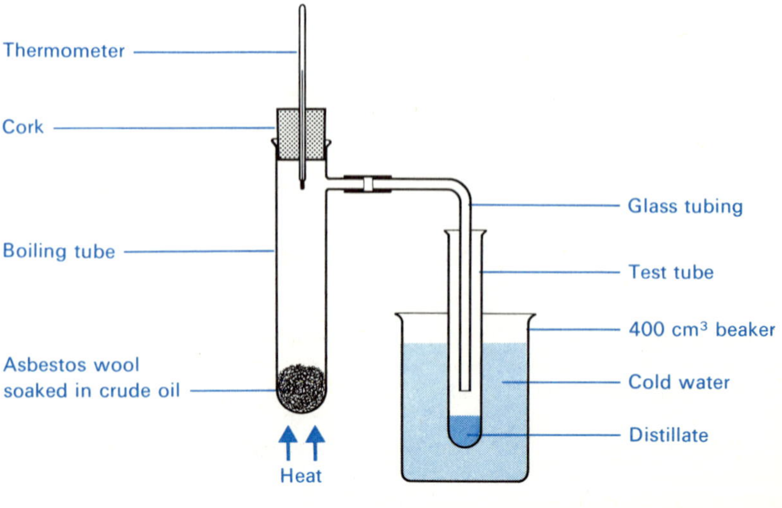

16 Explain briefly how you would separate each of the following mixtures. State also the property in which the components differ and upon which the method of separation is based.
(a) Two proteins dissolved in water (e.g., albumen and insulin in egg white).
(b) Petrol and water.
(c) Iron filings and powdered sulphur.
(d) Heptane (b. pt. 98°C) and octane (b. pt. 126°C), both of which are constituents of petrol.
(e) A fine suspension of chalk in water.
(f) Sugar and sand.
(g) Sawdust and soil.
(h) A grease stain from a cloth.

17 Vinegar is a dilute solution of acetic acid in water with small amounts of malt and spices added.
(a) Explain the two underlined words.
(b) What substance is the solute in vinegar?
(c) What substance is the solvent in vinegar?
(d) Suppose you are given a clear solution of vinegar. How would you show that it is neither pure water nor pure acetic acid?

18 Make a list of separation processes which are important:
(a) in industry.
(b) in medicine.
(c) in the home.

19 Figure 2.15 is part of an article entitled *Petroleum*. Read it through very carefully before answering these questions.
(a) Define the following terms:
Mixture Fraction Distillation
(b) What do you understand by the following?
Refining Condensing Hydrocarbon
(c) Draw a fully labelled diagram of the apparatus you would use to separate petrol (boiling range: 60°C–80°C) from crude oil in the laboratory.
(d) Name two other products of everyday importance obtained by the distillation of crude oil.
(e) Which of the following processes occur during distillation?
Chemical change Change of state
Evaporation Absorption of heat
Condensation Evolution of heat
(f) What do you think is the connection between molecular size and boiling point?

Figure 2.14

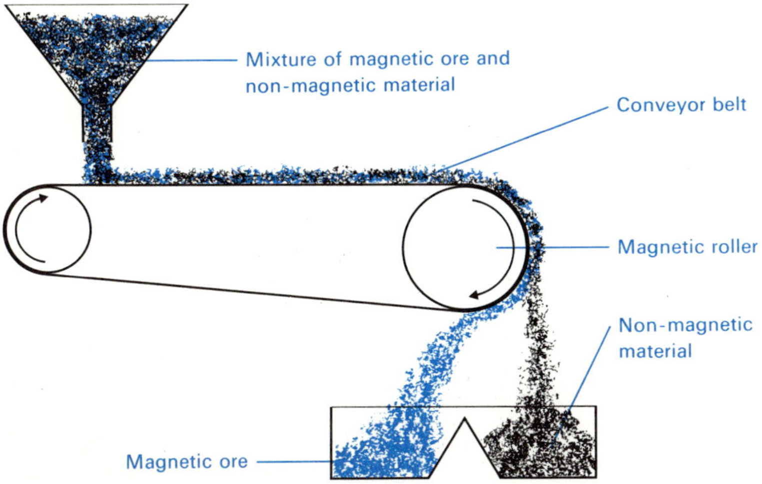

Figure 2.15

Crude oil is a complex mixture of hydrocarbons which is of little use as it comes out of the ground. The first stage in the refining process is to separate the hydrocarbon mixture into fractions whose constituents boil at a similar temperature – that is, into fractions which contain molecules of similar size. It is rather like separating a heap of coal into lumps of similar size by sifting the coal through sieves of finer and finer mesh. This separation is carried out by distilling the crude oil. You have come across distillation several times in the laboratory. Distillation in an oil refinery is essentially similar except that the liquid hydrocarbon mixture is separated not by boiling off the different fractions one after the other but by vaporizing nearly all the crude oil and collecting the separated fractions as the vapours condense back into liquids. Vaporized crude oil is passed into a distillation or fractionating column (which, in a typical oil refinery, may be as high as 70 m), and the hydrocarbon fractions, boiling at different temperatures and therefore at different heights in the column, are separated and collected.

No matter what you do, you cannot re-form bread from burnt toast.

30 Figure 3.1

Substance heated	Appearance before heating	Effect of heating	Effect of cooling	Weight change
Copper	Shiny pink metal	Becomes red hot	Shine lost, dull black coating	Gain
Baking powder	White powder	None visible	None visible	Loss
Zinc oxide	White powder	Turns yellow	White powder	None
Salt	White crystals	None visible	None visible	None
Red lead	Orange-red powder	Darkens	Becomes a yellow powder	Loss
Sulphur	Yellow crystals	Melts to an orange liquid, which darkens to red and then black. An orange vapour is given off	Yellow solid re-forms	Loss
Red mercury oxide	Red powder	Silvery liquid collects around top of test tube	Silvery liquid remains, red solid left in test tube	Loss
Copper sulphate crystals	Blue crystals	Turns white, moisture on sides of test tube	Stays white, moisture on sides of test tube	Loss
Magnesium	Shiny metal	Becomes coated with white powder	White flaky coating	Gain
Sand (silicon oxide)	Fine brown powder	None visible	None visible	None
Iodine	Purple-black crystals	Purple vapour forms	Purple-black solid forms on sides of test tube	Loss
Water	Clear liquid	Liquid boils, steam forms	Liquid re-forms on cooler sides of test tube	Loss

3 Action of heat on materials

3.1 Introduction

In chapter 2 we saw how to separate mixtures into pure substances. These pure substances contain only one kind of material and are simpler than the mixtures we started with.

Is it possible to obtain even simpler substances from these pure materials? For example, can we split up pure salt to get simpler substances? This would involve changing the salt into different substances; but we find that the methods described in chapter 2 will not split salt into different substances. Processes of filtration, distillation, floatation and chromatography *never change* substances into different ones. They can *only separate* one substance from another.

Try to answer these questions

1 Can you think of one process which can change a substance into different substances?
2 What happens when bread is toasted?
3 What happens if the bread is left toasting for 1 hour?
4 Have the substances in bread been changed into different substances?
5 What is the name of the black solid which forms when toast is burnt?

Coal also changes into different substances when it burns. Can you think of others? The properties of carbon, the black solid which forms when toast is burnt, are very different from those of bread. A different substance has been produced by heating bread. We call changes which result in the formation of new substances **chemical changes** or **chemical reactions.** The new substance formed during the chemical reaction is called the **product** of the reaction.

The chemical reaction which produced a black solid from bread is caused by heat energy from the cooker. Many of the foods we eat are prepared by chemical reactions caused by heat energy.

Figure 3.1 shows the effect of heating some substances in a hard-glass test tube with a bunsen burner, gently at first and then more strongly.

3.2 Changes in appearance on heating

Sometimes it is possible to see the changes which occur when substances are heated.

Look at the results in Figure 3.1 for:
1 changes of state, i.e., solids changing to liquids or liquids changing to gases, and vice versa.
2 changes in colour.
3 temporary changes, i.e., substances changing on heating but re-forming the original substance on cooling.
4 permanent changes, i.e., substances changing on heating but not re-forming the original substance on cooling.

MELTING, BOILING AND SUBLIMATION

Sulphur, iodine and water all show changes of state on heating.

The solid yellow sulphur crystals melt to an orange liquid at 119°C. Above 119°C, the sulphur is a liquid, below 119°C it is a solid. If sulphur liquid is heated beyond 119°C, it darkens to red and then to black. At 445°C the liquid sulphur boils to form an orange vapour. We can summarize what has happened by writing an **equation.** Chemists use equations to summarize their experiments in the same way that mathematicians use equations to

Sulphur melting

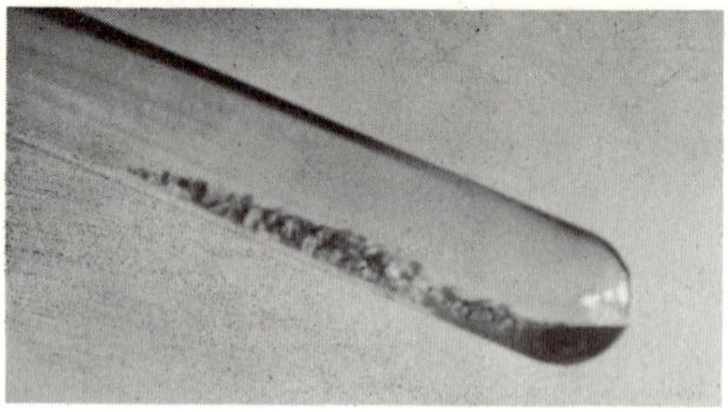

Solid carbon dioxide (*'dry ice'*) sublimes, leaving no messy liquid. Here it is used to keep polio vaccine frozen solid in transit. Deep frozen, the vaccine has a shelf life of 1 year; it lasts for a mere 30 days in an ordinary refrigerator

Figure 3.2

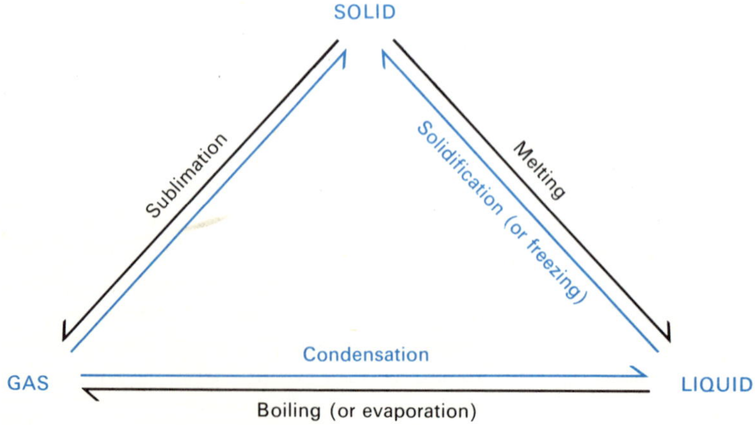

summarize their calculations. The equation summarizing what happens when sulphur is heated is:

$$\text{sulphur}_{(s)} \longrightarrow \text{sulphur}_{(l)} \longrightarrow \text{sulphur}_{(g)}$$

The small letters in brackets after the name of a substance indicate the state of the substance. They are called **state symbols.** $_{(s)}$ indicates a solid, $_{(l)}$ indicates a liquid, $_{(g)}$ indicates a gas or vapour and $_{(aq)}$ indicates an aqueous solution, i.e., a solution in water.

It is possible to include further information in the equation if necessary. A more detailed equation for the action of heat on sulphur is:

$\text{sulphur}_{(s)}$ *(yellow crystals)*

↓ *melts at 119°C*

$\text{sulphur}_{(l)}$ *(orange, red, then black liquid)*

↓ *boils at 445°C*

$\text{sulphur}_{(g)}$ *(orange vapour)*

On cooling the reverse process occurs, and solid crystals of sulphur will form when the temperature falls to 119°C.

Try to answer these questions

1 What is the melting point of ice?
2 What is the freezing point of water?
3 What is the boiling point of water?
4 In an equation, ice would be written as $\text{water}_{(s)}$. How would you write water and water vapour in an equation?
5 Write an equation for the change of state which occurs when water freezes to form ice.

When purple-black solid iodine is heated quickly it changes straight to a purple vapour without forming a liquid. The equation for the change is:

$$\text{iodine}_{(s)} \longrightarrow \text{iodine}_{(g)}$$

This change of state, from solid to vapour without passing through the liquid state, is called **sublimation.** We say that the iodine sublimes. We can purify some solids by sublimation. Solid iodine is obtained from the impure material by heating impure iodine in a brick furnace. The iodine is vaporized and solidifies on the cool surface of the metal chimney.

The changes of state which take place when substances are heated or cooled can be summarized in a cycle (Figure 3.2). Sometimes there is a loss of weight on heating because of the escape of vapour into the air.

TEMPORARY CHANGES

There are other changes besides melting and boiling in which the original substance re-forms on cooling. Zinc oxide is a white solid at room temperature. It changes colour to yellow on heating but re-forms the original white solid on cooling. An equation for this temporary change is:

$$\underset{(white)}{\text{zinc oxide}_{(s)}} \underset{cool}{\overset{heat}{\rightleftharpoons}} \underset{(yellow)}{\text{zinc oxide}_{(s)}}$$

Sulphur, iodine and water also undergo temporary changes of state on heating.

PERMANENT CHANGES

Look back at Figure 3.1. Which substances are permanently changed on heating? A permanent change results in the formation of a new substance. You can see a permanent new substance being formed when copper, red lead, red mercury oxide, copper sulphate crystals or magnesium are heated.

Heat turns blue copper sulphate crystals white and a vapour is given off which condenses to a colourless liquid on the cooler parts of the test tube.

Tests show that this liquid boils at 100°C and freezes at 0°C. This means the liquid could be water. Other liquids may also boil at 100°C and solidify at 0°C, so we should carry out further tests to show it is water. What tests would you make?

Adding water to the white solid re-forms the blue substance, and the test tube gets hot. The white solid is called anhydrous copper sulphate. 'Anhydrous' means 'without water'. The equation for this process is:

$$\underset{(blue)}{\text{copper sulphate crystals}_{(s)}} \xrightarrow{heat} \underset{(white)}{\text{anhydrous copper sulphate}_{(s)}} + \text{water}_{(l)}$$

Water is the only substance which reacts with anhydrous copper sulphate to form blue copper sulphate. Heat is given off in the reaction. Any liquid which contains water will turn anhydrous copper sulphate blue. Consequently, this reaction is a good test for the presence of water.

Copper sulphate crystals split up on heating to form anhydrous copper sulphate and water. A reaction in which one substance splits up into two or more simpler substances is called **decomposition.** The decomposition in this case was caused by heating, and so it is an example of **thermal decomposition.** ('Thermal' comes from the Greek word *therme*, meaning heat.) The solid left behind when a substance is heated is called the **residue.** In this experiment, the residue is white anhydrous copper sulphate.

When blue copper sulphate crystals are heated they break up (decompose) into white anhydrous copper sulphate and water. However, when the water is added back to the white solid, blue copper sulphate is re-formed. The simpler substances, water and anhydrous copper sulphate, can join together to form the more complex substance, blue copper sulphate. A reaction in which two or more substances join together to form another substance is called **synthesis.**

Synthesis is just the opposite of decomposition. Synthesis is a 'building up', whereas decomposition is a 'breaking down'. Thus:

$$\underset{(blue)}{\text{copper sulphate crystals}_{(s)}} \underset{synthesis}{\overset{decomposition}{\rightleftharpoons}} \underset{(white)}{\text{anhydrous copper sulphate}_{(s)}} + \text{water}_{(l)}$$

A reaction which can be reversed in this way by making the products react together to re-form the original substance is called a **reversible reaction.** Reactions which cannot be made to go in the reverse direction are **irreversible reactions.** The burning of toast is an irreversible reaction. No matter how the black solid and the smoke produced when the toast is burnt are treated, it is impossible to re-form the bread.

Try to answer these questions

Look carefully at Figure 3.1.

1 Which substances did not appear to change on heating?

2 Can there have been any change in a substance if it looks the same before, during and after heating?

3 Substance *X* loses weight on heating. Has it undergone a chemical change? Explain.

3.3 Changes in weight on heating

Sometimes it is impossible to see whether any change has taken place when a substance is heated. Neither baking powder, salt nor sand appears to change on heating. No changes in these solids are visible during heating or on cooling. But it is possible, for example, that salt may be reacting to form a second white solid very similar to the salt itself. A change may take place which cannot be seen. We can find out if anything has happened by weighing the substance before and after heating. Look carefully at the weight changes which occur when baking powder, salt and sand are heated.

Baking powder (sodium hydrogen carbonate) loses weight on heating but salt and sand do not (see Figure 3.3). When baking powder is heated it decomposes to form a different white solid (sodium carbonate), water and carbon dioxide. If the baking powder is heated carefully, drops of water collect on the cooler parts of the test tube and carbon dioxide escapes into the air. The equation is:

$$\text{baking powder}_{(s)} \longrightarrow \text{sodium carbonate}_{(s)} + \text{water}_{(l)} + \text{carbon dioxide}_{(g)}$$

A simple test for carbon dioxide is to bubble the gases produced into lime water (see Figure 3.4). The lime water turns milky if carbon dioxide is present.

Thus we can tell from weight changes whether the substance produced on heating is more simple, more complex or the same as the starting material.

Figure 3.3

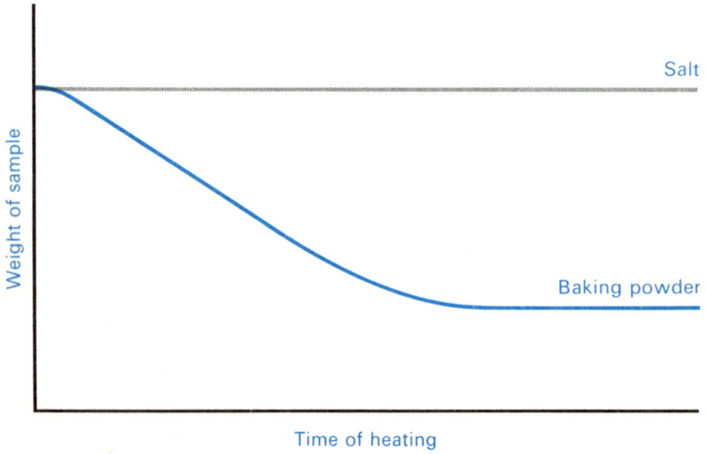

The substances, tested in this way, fall into three classes:

1 Substances which do not change in weight on heating such as salt and sand. These substances do not react on heating.

2 Substances which gain weight on heating, such as copper and magnesium. These substances react on heating and since the product weighs more than the starting material, it is probable that synthesis has taken place to form a more complex substance.

3 Substances which lose weight on heating such as baking powder, red lead, sulphur, red mercury oxide, blue copper sulphate, iodine and water. Three of these substances, sulphur, iodine and water, undergo changes of state on heating, and the loss in weight is due to the escape of vapour of the substance. On cooling the same starting material re-forms, but there is less of it.

Baking powder, red lead, red mercury oxide and blue copper sulphate all lose weight on heating and new substances form. The new substances are different colours from the starting materials, except in the case of baking powder. Since the products weigh less than the starting materials, decomposition has taken place. Why do some substances lose weight on heating whereas others gain in weight?

3.4 Investigating the loss in weight

Try to answer these questions

1 What substances are produced when blue copper sulphate is heated?

2 What are the states of these products?

3 Will any of the products escape from the test tube on heating?

4 Why does blue copper sulphate lose weight on heating?

When baking powder is heated, carbon dioxide and water vapour are produced which escape from the open end of the test tube. As a result of these gases escaping there is a loss in weight on heating. Red lead and red mercury oxide may also lose weight on heating because gases escape into the air.

Figure 3.5 shows a possible method of collecting any gases produced when a substance is heated. As the gas collects the water is pushed out of the inverted test tube. This

Figure 3.4

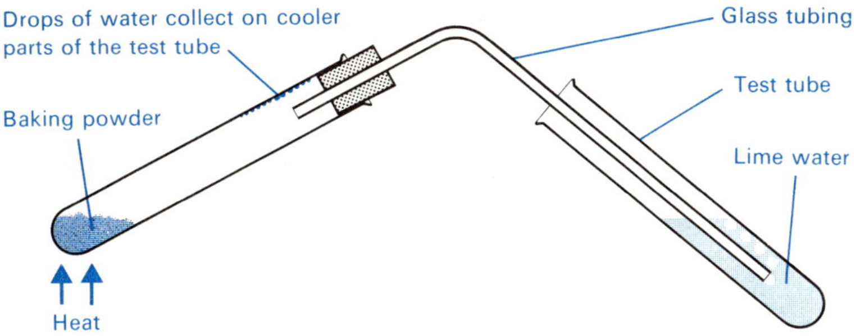

Figure 3.5

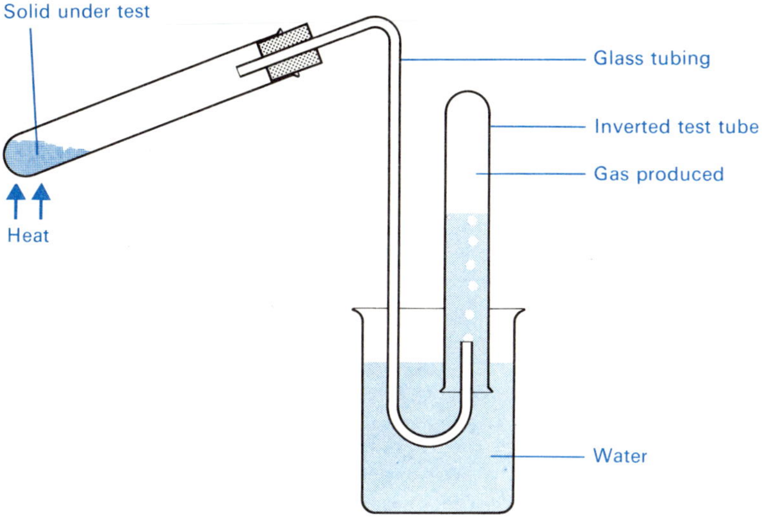

Figure 3.6

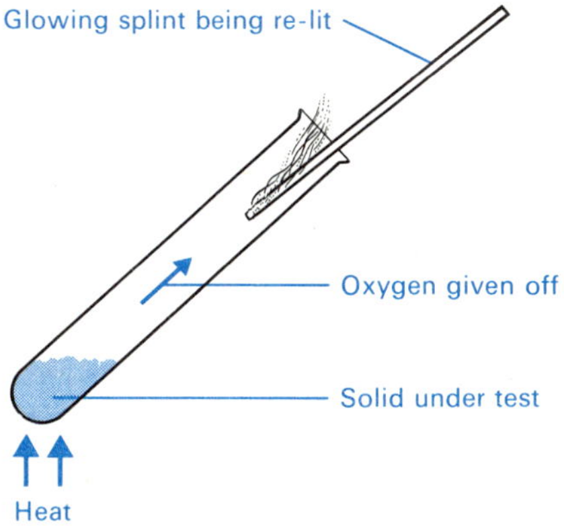

method of collecting a gas is called collection over water. It is only suitable for collecting gases that do not react with, or dissolve in, water. Why could you not collect a water-soluble gas this way?

When sugar is heated in the apparatus in Figure 3.5, a colourless liquid condenses on the cooler parts of the hard-glass test tube and a mixture of gases and vapours collects in the inverted test tube. A black solid remains in the heating tube.

The colourless liquid turns anhydrous copper sulphate blue, showing that it contains water. The gases turn lime water milky, showing that carbon dioxide is also formed. The black solid left in the heating tube is charcoal (carbon). The loss of water and carbon dioxide from the sugar will cause a loss in weight on heating. An equation for the reaction is:

$$\underset{(white)}{\text{sugar}_{(s)}} \xrightarrow{heat} \underset{(black)}{\text{carbon}_{(s)}} + \text{water}_{(l)} + \text{carbon dioxide}_{(g)}$$

When red lead and red mercury oxide are heated in the apparatus in Figure 3.5, no water condenses on the cooler parts of the hard-glass test tube in either case, nor is there any carbon dioxide in the gases produced.

However, a colourless gas collects in the inverted tube when either solid is heated.

In the case of red mercury oxide a silvery liquid coats the top of the heating tube and the gas collected will re-light a glowing splint (i.e., a lighted splint whose flame has been blown out, leaving a glowing ember). The silvery liquid is mercury and the gas produced is oxygen. A simple test for oxygen is that it re-lights a glowing splint. Figure 3.6 shows how to test for oxygen with a glowing splint.

An equation for the thermal decomposition of red mercury oxide is:

$$\underset{(red)}{\text{mercury oxide}_{(s)}} \xrightarrow{heat} \underset{(silvery)}{\text{mercury}_{(l)}} + \underset{(colourless)}{\text{oxygen}_{(g)}}$$

The loss of oxygen from the test tube on heating explains the loss in weight. In the case of red lead, the yellow solid left behind is lead oxide, and the colourless gas collected over water is oxygen. What causes the loss in weight?

$$\underset{(red)}{\text{red lead}_{(s)}} \xrightarrow{heat} \underset{(yellow)}{\text{lead oxide}_{(s)}} + \underset{(colourless)}{\text{oxygen}_{(g)}}$$

3.5 Investigating the gain in weight

Try to answer these questions

When copper is heated strongly it becomes dull black and increases in weight.

1 Has the copper lost or gained something during heating?

2 Could decomposition have occurred during heating?

3 Could synthesis have occurred during heating?

4 Where has the increase in weight come from?

The increase in weight could not have come from the bunsen burner flame. The flame did not touch the copper. If the bunsen flame could cause an increase in weight, even an empty test tube would gain weight on heating.

When copper is heated in a glass container from which all the air has been removed (an evacuated container) there is no increase in weight. Thus, the increase in weight has come from the air. Which gas in the air has reacted with the copper? Which gas was reactive enough to re-light a glowing splint? – Oxygen. When copper is heated strongly it reacts with the oxygen in the air to form a coating of black copper oxide on its surface. The product is more complex than either of the starting materials, copper or oxygen:

$$\text{copper}_{(s)} + \text{oxygen}_{(g)} \xrightarrow{\textit{heat}} \underset{(\textit{black})}{\text{copper oxide}_{(s)}}$$

This is an example of synthesis. The increase in weight of the copper is due to its reaction with oxygen.

Magnesium ribbon burns with a brilliant flame

A similar reaction accounts for the increase in weight of magnesium on heating. In this case magnesium oxide is formed:

$$\text{magnesium}_{(s)} + \text{oxygen}_{(g)} \xrightarrow{\textit{heat}} \underset{(\textit{white})}{\text{magnesium oxide}_{(s)}}$$

The products of reaction with oxygen are called **oxides.**

One striking difference between the reaction of air with magnesium and with copper is the difference in the speeds with which the reactions occur. Magnesium reacts so vigorously with oxygen that it will burn in air with a very bright white flame. Once the magnesium has been given enough heat to start the reaction, so much heat is produced that the reaction can continue without any further heating.

With copper, the reaction is very slow and the copper only reacts if it has a continuous supply of heat. Sodium reacts with the oxygen in air even more rapidly than magnesium. Sodium is a soft metal. A freshly cut surface becomes covered with a layer of sodium oxide in a few seconds even without heating:

$$\text{sodium}_{(s)} + \text{oxygen}_{(g)} \longrightarrow \text{sodium oxide}_{(s)}$$

Because of this rapid reaction with oxygen in the air, sodium is stored under oil. (It cannot be kept under water, since it reacts vigorously with water as well as oxygen). Copper, magnesium and sodium all react in a similar way with oxygen to form metal oxides, but the speeds of their reactions are very different. We say that their **rates of reaction** are different.

Some substances (such as sodium) will react without being heated. Other chemical processes which occur without heating are the setting of cement, the souring of milk and the rusting of iron. These processes take place much more slowly than the reaction between freshly cut sodium and oxygen.

Some substances (such as magnesium) need heat to start a reaction, but once the reactions are 'triggered off' they can continue without further heating. Lighting a bunsen burner is an example of this sort of reaction. Gas will not burn until the heat of a lighted match 'triggers off' a reaction.

Other substances (such as copper) only react if they are heated all the time. Once the heating stops, the reaction also stops.

Summary

1 A chemical reaction occurs when a new substance is produced. The new substance is called the product of the reaction.

2 A chemical equation summarizes what happens in a chemical reaction.

3 State symbols are used in equations to indicate whether a substance is solid $_{(s)}$, liquid $_{(l)}$, gas $_{(g)}$ or a solution in water $_{(aq)}$.

4 Sublimation is the change of state from solid to vapour without passing through the liquid state.

5 Decomposition occurs when one substance splits up into two or more simpler substances.

6 Synthesis occurs when two or more substances join together to form a single more complex substance. Synthesis is the reverse of decomposition:

$$A + B \underset{decomposition}{\overset{synthesis}{\rightleftharpoons}} C$$

7 A reversible reaction is one which will go in either the forward or the reverse direction. An irreversible reaction cannot be made to go in the reverse direction.

8 The three types of change which may occur when substances are heated are:
(a) change of state.
(b) synthesis (substances join together).
(c) decomposition (substances are broken apart).

9 When substances are heated in air the results fall into three categories:
(a) Some substances do not change in weight on heating. They do not react on heating. Solids, however, may melt on heating, but re-form on cooling.
(b) Some gain weight on heating. These react with oxygen in the air to form solid oxides.
(c) Some lose weight on heating for one of three reasons:
(i) They decompose and lose gases such as water vapour or carbon dioxide to the air.
(ii) They change to a gas without decomposing (as in boiling), and the gas escapes from the test tube.
(iii) They react with oxygen in the air to form gaseous oxides which escape into the air.

Study questions

1 Purple crystals of potassium permanganate were heated in a test tube. A gas was produced during the heating.
(a) How would you find out if the gas produced is:
(i) carbon dioxide.
(ii) water vapour.
(iii) oxygen?
(b) What experiments would you carry out to see if this reaction is reversible?
(c) Suppose oxygen is produced when potassium permanganate is heated.
(i) Is the reaction an example of synthesis?
(ii) Is the reaction an example of decomposition?
Explain your answers.
(d) How would you show that your ideas in part (c) are correct?

2 Some finely powdered solid was heated gently and constantly for 20 min. The temperature was recorded every 2 min. Here are the results:

Time (min)	0	2	4	6	8	10	12	14	16	18	20
Temp (°C)	15	27	38	47	51	51	51	56	64	71	76

(a) How would you ensure that the heating of the container was evenly distributed?
(b) Draw a graph of the results, plotting time along the horizontal axis and temperature along the vertical axis.
(c) What is the temperature of the solid after 5 min?
(d) After how many minutes is the temperature 60°C?
(e) How long did it take for the temperature to rise from 20°C to 40°C?
(f) How long did it take for the temperature to rise from 55°C to 75°C?
(g) Why are these two periods, for the same temperature rise, not the same?
(h) What is the melting point of the solid? Is the solid pure?

3 (a) A red powder, *A*, darkened on heating and produced a colourless gas, *B*, which re-lit a glowing splint, and beads of a shiny metal, *C*, which did not solidify when cooled to room temperature.
(b) *B* reacted with copper metal to give a black solid, *D*.
(c) When *B* is passed over a hot black

solid, *E*, a colourless gas, *F*, was synthesized which turned lime water milky.
(d) *D* and *F* are both produced when the green powder, *G*, decomposes on heating.
Identify the substances *A* to *G*. Write equations for reactions (a), (b), (c) and (d) with state symbols. Explain what you understand by the words underlined.

4 When a piece of copper is heated in a bunsen flame it turns black. Four explanations were suggested for what happened:
(a) Soot from the flame has been deposited on the copper.
(b) The copper has combined with something from the air.
(c) The heat of the flame has driven something out of the copper.
(d) Copper has combined with something in the flame.
What experiments would you carry out to test each explanation? Say what the results of your tests would be.

5 A piece of soap in the shape of a cat develops a furry appearance when exposed to the air for a few days. What experiments would you carry out to test whether this is due to:
(a) loss of water from the soap.
(b) water vapour in the air.
(c) oxygen in the air.
(d) carbon dioxide in the air?

6 Draw a labelled diagram of the apparatus you would use to collect the water produced when pink crystals of cobalt chloride are heated. What tests would you make to determine whether the liquid collected is pure water? How would you test whether the reaction is reversible?

7 Lime water is an aqueous solution of calcium hydroxide. When carbon dioxide is bubbled into lime water, it goes milky. The milkiness is due to the formation of solid calcium carbonate in suspension.
(a) What is the solvent in lime water?
(b) What is the solute in lime water?
(c) Write a word equation for the reaction of carbon dioxide with lime water.
(d) How would you obtain a dry sample of the calcium carbonate formed when carbon dioxide is bubbled into lime water?
(e) Draw a fully labelled diagram of the apparatus you would use to discover whether there is more carbon dioxide in exhaled air than in inhaled air?

8 Figure 3.7 is the record of a student who had been finding out what happened when various substances were heated in air.
(a) Which substance is not changed on heating?
(b) Which substances are decomposed on heating?
(c) Which substances are decomposed on

Figure 3.7

Substance heated	Appearance before heating	Effect of heating	Effect of cooling	Weight change	Effect of adding water to the cold solid after heating
Cobalt chloride	Pink crystals	Turns blue, moisture forms on sides of test tube	Solid remains blue	Loss	Pink colour returns, solid dissolves to form a pink solution
Bread	Soft white solid	Turns brown, then black and powdery	Black powder remains	Loss	Does not dissolve
Copper carbonate	Green powder	Turns black	Black solid remains	Loss	Does not dissolve
Potassium chloride	White crystals	No change	No change	None	Dissolves to give a colourless solution
Iron wire	Shiny metal	Becomes red hot	The shine is lost and the surface is grey-black	Gain	Does not dissolve

heating, but possibly re-form when water is added to the solid residue after heating?

(d) What substance may be responsible for the loss in weight when cobalt chloride was heated?

(e) The solid residue on heating pink crystals of cobalt chloride is anhydrous cobalt chloride. Write an equation for the reaction with state symbols.

(f) What substances may be responsible for the loss in weight when bread was heated? What is the name of the black powder which remains?

(g) Is it possible to tell from these experiments whether the following are soluble or insoluble in water?

(i) Copper carbonate.
(ii) Potassium chloride.
Explain your answer.

(h) Explain the increase in weight of iron wire. Write an equation for the reaction with state symbols.

9 Three separate pieces of magnesium are heated in:

An open flask.
A sealed flask.
An evacuated flask.

(a) Write a word equation for the reaction of magnesium when heated in air.

(b) Will the open flask gain weight, lose weight or stay the same weight on heating? Explain.

(c) Will there be any chemical reaction in the sealed flask?

(d) Why does the sealed flask not increase in weight on heating?

(e) Will there be any chemical reaction in the evacuated flask? Explain.

(f) Will anything at all happen to the magnesium in the evacuated flask on heating?

10 Some glues are thought to react with the air during the setting process. Devise a method of testing whether Bostik glue reacts with the air when it sets.

11 Some baking powder was heated in the apparatus shown in Figure 3.5. Several test tubes of gas were collected before the bubbling ceased. The gas produced from the baking powder is carbon dioxide.

(a) How would you confirm that the gas collected contains carbon dioxide?

(b) The first test tube of gas does not give the test for carbon dioxide but all the other tubes do. What gas does the first tube contain? Explain why the first tube contains no carbon dioxide.

(c) When the bubbling ceases the delivery tube is quickly detached from the heated tube, otherwise water would be sucked back along the delivery tube into the heated tube. Explain why the water gets sucked back in this way.

12 A sample of green nickel carbonate, heated in a closed flask containing air, decomposed to form black nickel oxide and carbon dioxide.

(a) Write an equation for the reaction.

(b) The closed flask is weighed before and after heating. Will there be any weight change on heating? Explain.

(c) The flask is opened and re-weighed. Will there be any change in weight? Explain.

13 A student heated a substance in a flask and noted that the flask and its contents increased in weight.

(a) Was the flask open or sealed?

(b) The student reasoned that heat was the only thing to have entered the flask and so he had weighed heat. Do you agree with him?

(c) Can you think of any other explanations for the increase in weight?

(d) Explain carefully how you would demonstrate to the student that his suggestion that heat alone had caused the increase in weight is incorrect.

14 When carbon is heated very strongly, it reacts with oxygen in the air to form carbon dioxide.

(a) Is carbon a solid, a liquid or a gas?

(b) Is carbon dioxide a solid, a liquid or a gas?

(c) Write an equation with state symbols to summarize the reaction of carbon with oxygen to produce carbon dioxide.

(d) Will the carbon dioxide produced weigh more or less than the carbon from which it is formed?

(e) Will the carbon dioxide, which is formed, stay in the test tube or escape as a gas?

(f) Will there be an increase or decrease in weight when carbon is strongly heated in an open tube?

These buildings are all built from the same few basic pieces

Figure 4.1

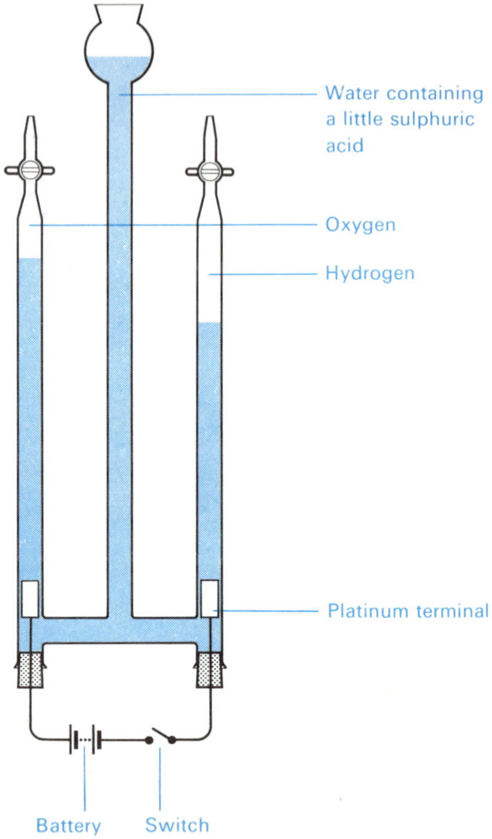

4 Elements – the building blocks

4.1 Introduction

The idea that all materials are built up from a few simple substances is not new. But it took scientists a long time to discover just what these simple substances were. We can use techniques such as filtration, distillation and chromatography to separate mixtures into pure substances. These pure substances are simpler than the starting mixtures.

In the last chapter, we discovered that some of these pure substances could be broken down by heating to form even simpler substances. Red mercury oxide splits up on heating to form shiny beads of mercury and oxygen. But we cannot decompose mercury or oxygen any further. This suggests that there must be some pure substances which cannot be broken down into simpler substances. In 1661, Robert Boyle suggested the name **element** for a substance which could not be broken down into a simpler substance. This is still a fairly satisfactory definition of an element, more than 300 years later.

4.2 Reactions which produce elements

Heating red mercury oxide yields two substances, mercury and oxygen. Neither can be broken down to a simpler substance:

$$\underset{(red)}{\text{mercury oxide}_{(s)}} \xrightarrow{heat} \text{mercury}_{(l)} + \text{oxygen}_{(g)}$$

Therefore, mercury and oxygen are elements.

When sugar is heated a black solid, carbon, is left behind. Carbon cannot be broken down into any simpler substance. Therefore, carbon is an element.

$$\text{sugar}_{(s)} \xrightarrow{heat} \text{carbon}_{(s)} + \text{carbon dioxide}_{(g)} + \text{water}_{(g)}$$

Try to answer these questions

When carbon is heated in oxygen it reacts to form carbon dioxide.

1 Is carbon dioxide an element?

2 Which elements does carbon dioxide contain?

3 Is carbon dioxide more or less complex than carbon?

4 What type of reaction is this?

5 Write an equation for the reaction.

Some substances which contain more than one element are not decomposed on heating. For example, if water is heated and then condensed the final product is exactly like the starting material. No chemical reaction has taken place. But when electricity is passed through water, hydrogen and oxygen are produced. We can use the apparatus in Figure 4.1 for this experiment. The water contains a little sulphuric acid which helps the water to decompose more easily. Bubbles of oxygen rise from one of the platinum terminals and bubbles of hydrogen rise from the other. Twice as much hydrogen as oxygen is produced.

$$\text{water}_{(l)} \xrightarrow{electricity} \text{hydrogen}_{(g)} + \text{oxygen}_{(g)}$$

The hydrogen and oxygen cannot be made any simpler; they are elements. This decomposition is brought about by electricity and is called **electrical decomposition,** or **electrolysis.**

The oxygen and hydrogen which collect above the platinum terminals can be transferred to test tubes by opening the taps. The tube containing hydrogen should be held upside-down all the time to stop the lighter-than-air hydrogen from floating out. The oxygen can be tested by showing that it re-lights a glowing splint.

If a lighted splint is placed in the inverted test tube of hydrogen a small explosion occurs and a 'pop' is heard. This 'pop' is used as a test for hydrogen.

4.3 How many elements are there?

Eight elements (copper, gold, silver, mercury, lead, iron, sulphur and carbon) have been known since prehistoric times. This is because they occur naturally (copper, gold, silver, sulphur and carbon) or they can be extracted easily from their ores (mercury, lead and iron).

This gold dagger handle is 2300 years old. The intricate patterns have not corroded with age

By 1800, only about 30 elements were known. Most elements were discovered in the nineteenth century, and by 1900 about 90 different elements had been found in the earth's crust or atmosphere. More recently, chemists have made new elements by changing one element into another in nuclear reactors. Many of these man-made elements, unlike other elements, are very unstable and they break up to form more stable elements. By 1974 there were 105 known elements, but it is likely that other man-made elements will be produced in due course.

At one time it was very difficult to be sure that a substance had been decomposed as much as possible. Sand (silicon oxide) is not decomposed by heat or by electricity. Yet, silicon oxide is not an element; it contains silicon and oxygen joined together. As a result, silicon oxide, and other substances which could not be decomposed, were wrongly classed as elements. Nowadays, chemists can use an instrument called a mass spectrometer to decide whether a substance contains one element or more.

Every substance which has been investigated can be split up into one or more of the 90 elements which exist in the atmosphere or in the earth's crust. All other substances can be built up from these 90 elements.

4.4 Classifying the elements

In chapter 1, we classified substances as solids, liquids and gases. Let's try to classify elements in this way. Some common elements are listed in Figure 4.2 classified as solids, liquids and gases at room temperature.

Look at the solids in Figure 4.2. Are they similar? Six of them (copper, gold, silver, lead, iron and magnesium) are shiny and can be hammered into shape (malleable). They are called **metals.** The other two (carbon and sulphur) do not have a shiny appearance and cannot be hammered into shape. These elements are called **non-metals.**

The two liquids are also dissimilar. Mercury is a shiny liquid metal with a very high density (13.6 g cm^{-3}), whereas bromine is an orange-brown non-metal with a density of 3.1 g cm^{-3} (density is the mass of 1 cm^3 of a substance). No pattern emerges from classifying elements

Figure 4.2

Solids	Liquids	Gases
Carbon	Bromine	Chlorine
Copper	Mercury	Hydrogen
Gold		Nitrogen
Iron		Oxygen
Lead		
Magnesium		
Silver		
Sulphur		

Figure 4.3

Element	Melting point in °C	Boiling point in °C	Density in g cm^{-3}
Carbon	3730	4830	2.3
Hydrogen	−259	−253	8×10^{-5}
Oxygen	−219	−183	13×10^{-4}
Sulphur	119	445	2.1
Copper	1083	2600	9.0
Iron	1540	3000	7.9
Magnesium	650	1110	1.7
Silver	961	2210	10.5

Figure 4.4

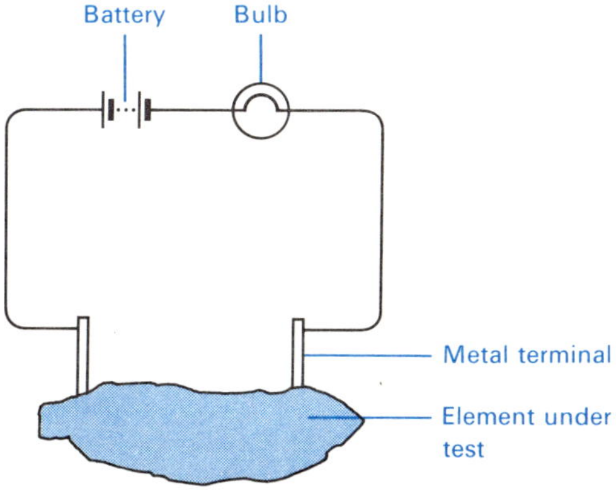

Figure 4.5

Property	Metals	Non-metals
State	Usually solids at room temp.	Solids, liquids or gases at room temp.
M. pt. and b. pt.	Usually high	Usually low
Density	Usually high	Usually low
Appearance	Shiny and lustrous	Dull without lustre
Effect of hammering (malleability)	Can be hammered into shape (malleable)	Brittle (non-malleable) and soft when solid
Thermal and electrical conductivity	Good	Poor

as solids, liquids and gases. Is there a better method of classifying elements? Yes, as metals and non-metals.

4.5 The properties of metals and non-metals

Try to answer these questions
Look at the properties of the elements listed in Figure 4.3.

1 Which elements have high density (greater than 5 g cm^{-3})?
2 Which elements have low density (less than 1 g cm^{-3})?
3 Which elements have high melting points (greater than 500°C)?
4 Which elements have high boiling points (greater than 1000°C)?
5 Which of the elements can be polished to a shine and have a lustre?
6 Which of the solid elements can be bent or hammered into different shapes?

Do these properties separate the metals from the non-metals? Are there any elements which resemble metals in some properties and non-metals in others? One simple experiment which helps us to tell clearly the difference between metals and non-metals is to see whether they conduct electricity. We can use the apparatus in Figure 4.4.

The bulb lights when the element tested is a metal or graphite (a form of carbon) showing that metals and graphite are good conductors of electricity. When non-metals (other than graphite) are tested, the lamp does not light, showing that non-metals (except graphite) are poor conductors of electricity.

Although there are one or two exceptions, such as graphite, the classification of elements as metals and non-metals is very useful. Most of the 105 known elements are metals. There are about 80 metals and about 20 non-metals. Figure 4.5 lists the general differences in properties between metals and non-metals.

Metals usually have high melting point, high boiling point and high density, a shiny surface and good conductivity, whereas non-metals usually have low melting point, low boiling point and low density, a dull surface and poor conductivity. Notice that, in general, metals show one pattern in behaviour and properties and non-metals show another.

Figure 4.6

Element	State at 20°C	Boiling point in °C	Density in g cm $^{-3}$	Thermal conductivity
Barium	s	1640	3.5	Good
Iodine	s	183	4.9	Poor
Krypton	g	−152	3.4×10^{-3}	Poor
Selenium	s	685	4.8	Poor
Strontium	s	1380	2.6	Good
Titanium	s	3260	4.5	Good

Graphite is such a good conductor of electricity that it is used as one of the electrodes in torch batteries

Figure 4.7

Metals	Non-metals
Aluminium	Argon
Calcium	Chlorine
Copper	Fluorine
Gold	Carbon
Iron	Helium
Lead	Hydrogen
Lithium	Neon
Magnesium	Nitrogen
Mercury	Oxygen
Potassium	Phosphorus
Silver	Silicon
Sodium	Sulphur
Tin	
Zinc	

Try to answer this question

Say whether you think the elements listed in Figure 4.6 are metals or non-metals.

Use a data book to look up the same properties for three other elements of your choice: classify your elements as metals or non-metals. Can you classify an element from its density?

Elements which are exceptions to the usual generalizations have some properties like metals and other properties like non-metals. For example, graphite resembles non-metals in having a low density and no lustre and in being brittle; but it resembles metals in having a high melting point, a high boiling point and good electrical conductivity. Mercury is unlike other metals because it is a liquid.

It is not always easy to classify an element as either metal or non-metal. What did you make of selenium in Figure 4.6? Other more suitable methods of classifying and arranging the elements will emerge later. Lists of metals and non-metals are shown in Figure 4.7.

4.6 Combining elements

Some substances, such as copper and magnesium, increase in weight when they are heated in air:

$$\text{copper}_{(s)} + \text{oxygen}_{(g)} \longrightarrow \text{black copper oxide}_{(s)}$$

$$\text{magnesium}_{(s)} + \text{oxygen}_{(g)} \longrightarrow \text{magnesium oxide}_{(s)}$$

The oxides are more complex than the starting materials; they cannot be elements since each contains two elements. The copper and oxygen have not simply mixed together. They have joined together to form a new substance, black copper oxide. We say that copper and oxygen have **reacted** or **combined,** and the black copper oxide is called a **compound.** Magnesium oxide is also a compound; it contains the elements magnesium and oxygen joined together. Sugar is a compound of the elements carbon, hydrogen and oxygen. A compound is a substance that contains two or more elements combined together.

Try to answer these questions

1 When red lead is heated it decomposes to oxygen and a yellow solid. Could red lead be

an element?
2 Could red lead be a compound?
3 What elements does red lead contain?

4.7 Differences between mixtures and compounds

When two or more elements combine together they form a compound whose properties differ from those of the combining elements. For example, black copper oxide is very different from the shiny pink metal, copper, and the colourless gas, oxygen. Hydrogen, a colourless gas, burns in air or oxygen to form hydrogen oxide (water), a colourless liquid:

$$\text{hydrogen}_{(g)} + \text{oxygen}_{(g)} \longrightarrow \text{hydrogen oxide (water)}_{(l)}$$

When hydrogen is burnt in air the water produced can be condensed to a liquid by allowing the flame to play on the side of a flask filled with cold water (Figure 4.8).

The product, water, has very different properties from the colourless gases from which it has come. When hydrogen and oxygen combine to form water a chemical reaction takes place and heat is produced in the flame. Reactions which give out heat are called **exothermic.**

But, when hydrogen and oxygen are just mixed together no chemical reaction takes place and heat is not produced. This is the first important difference between a pure compound and a mixture. A chemical reaction takes place when a compound is formed; the compound is a new substance. There is no chemical reaction when a mixture is formed; there is no new substance. The reactions between elements to form compounds are usually exothermic. Can you think of two examples of exothermic reactions when elements combine?

The properties of a mixture of hydrogen and oxygen are between those of pure hydrogen and pure oxygen. The properties of water, a compound of hydrogen and oxygen, are very different to those of pure hydrogen and pure oxygen. The properties of hydrogen and oxygen in water are changed by the combination. This is the second important difference between a pure compound and a mixture. The properties of a pure compound differ from those of the constituent elements, but the properties of a mixture are in between those of its constituents.

When iron filings are mixed with powdered sulphur no chemical reaction occurs. The iron filings can be separated from the mixture very easily with a magnet. The sulphur can be separated from the mixture by dissolving it in carbon disulphide, in which iron is insoluble. However, if the mixture is heated a chemical reaction takes place between iron and sulphur, forming iron sulphide, a dark grey solid. Once

Figure 4.8

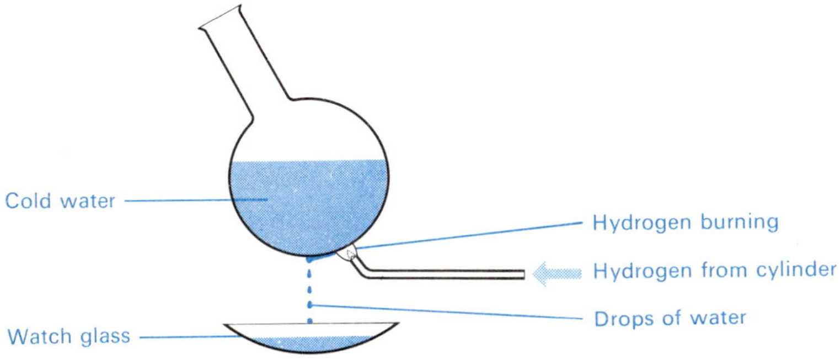

Iron and sulphur mixtures and (*far right*) iron sulphide, a compound containing 36.5% sulphur

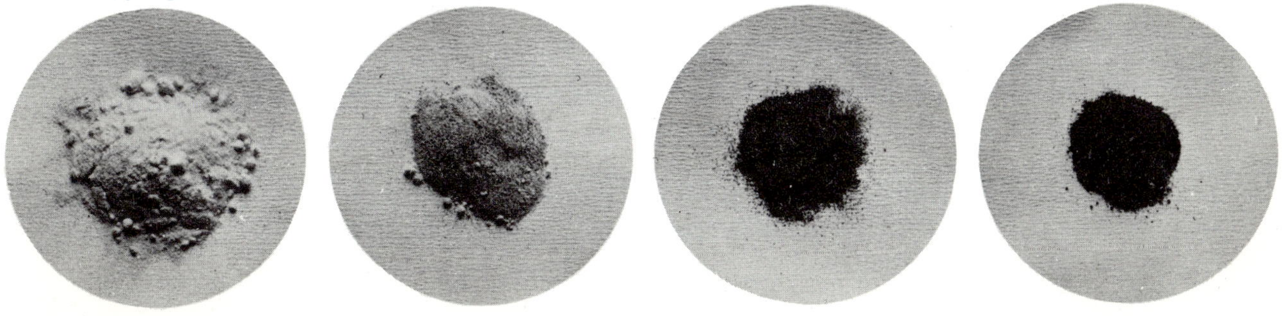

the reaction has started it will continue of its own accord, since the reaction is exothermic. The solid glows red hot during the reaction:

$$\text{iron}_{(s)} + \text{sulphur}_{(s)} \longrightarrow \text{iron sulphide}_{(s)}$$

The iron sulphide is very different from the starting materials, iron and sulphur. In iron sulphide the iron is firmly joined (bonded) to the sulphur and cannot be separated from it without difficulty. This is the third important difference between pure compounds and mixtures. In a mixture the components can usually be separated from each other fairly easily by one of the methods in chapter 2. The elements in a compound can only be separated after a further chemical reaction. A mixture of iron and sulphur is easily separated, but iron and sulphur can only be separated from each other in iron sulphide by a series of chemical reactions.

Suppose you were asked to make 100 g of a mixture of iron filings and sulphur. In order to make such a mixture you could take any weights of iron and sulphur which together weighed 100 g. For instance, you could mix 99 g of iron with 1 g of sulphur or 1 g of iron and 99 g of sulphur. Thus, the mixture of iron and sulphur could have any composition.

If you were asked to make 100 g of the compound, iron sulphide, you would have to heat a mixture of iron filings and powdered sulphur to a high temperature in order to make them react and form iron sulphide. If you analysed the iron sulphide you would find that it always contained about 63.5% iron and 36.5% sulphur. Pure iron sulphide always has the same composition. This is also true of other compounds. Water always contains 11.1% hydrogen and 88.9% oxygen. Black copper oxide always contains 80% copper and 20% oxygen.

The fact that pure compounds always have the same composition was stated by Proust as long ago as 1799 in the **law of constant composition**:

All samples of the same pure compound contain the same elements combined together in the same proportions by weight.

This is the fourth important difference between pure compounds and mixtures. The composition of a mixture can vary, but that of a compound is always the same. Nowadays chemists can analyse substances extremely accurately and detect very small variations in composition. Small differences in the composition of some compounds have been detected by chemists, but, in general, the composition of most compounds is always the same. Figure 4.9 summarizes the main differences between pure compounds and mixtures.

The difference between mixing and

Figure 4.9

	Compounds	Mixtures
1	A new substance is formed and a chemical reaction takes place when the compound forms	No new substance is formed and no chemical reaction takes place when the mixture forms
2	The properties are very different to those of the constituent elements	The properties are in between those of the constituents
3	The component elements can only be separated by chemical reaction	The component substances can be separated easily
4	The composition is always the same	The composition can vary

combining is shown by the following example. When you are running about and playing with other children, you are mixing with them but, if you hold hands firmly and do things together you are combining with them. In the same way, chemical combining involves a firm link or bond between one element and another.

Try to answer these questions
Classify the following substances as pure compounds or mixtures:

Air	Sea water	Sodium chloride
Soil	Milk	Distilled water
Sugar	Paint	Soap powder

Are you unsure about classifying any of these substances? In order to decide whether a substance is a pure mixture or a compound, ask yourself questions about the differences listed in Figure 4.9. Let's consider soil in this way.

1 Is any new chemical produced when soil forms? If all the components in soil (sand, grit, humus, moisture, etc.) are mixed together no new substance forms and no chemical reaction takes place. Soil is probably a mixture.

2 Are the properties of soil very different from the properties of its constituents or in between them? The colour of soil depends on the colour of the constituents. If the soil is very sandy, the colour of the soil will be yellowish. The dampness of the soil depends on the water content. This is further evidence that soil is a mixture.

3 Can the component parts in the soil be separated easily? Moisture in the soil can easily be separated by heating the soil to drive off water vapour and then condensing this to water. The grit could be separated fairly easily with a large sieve; more evidence that soil is a mixture.

4 Is the composition of soil always the same or does it vary? The composition of the soil varies greatly from one area to another. In some areas it is very chalky; in other areas, sandy. All the evidence suggests that soil is a mixture.

Ask yourself similar questions about the other substances.

4.8 The uses of metals and non-metals

The uses of an element depend on what it will or will not do, and on how easily it can be obtained. Metals, such as gold and silver, and the non-metal, carbon (in the form of diamond), which are attractive and unreactive are used to make ornaments and jewellery. Copper is a good electrical conductor, it is cheap compared to most metals and can be drawn into wires. Thus it is used in making electrical goods, wires and cables. Lead, a soft malleable metal which does not rust, is useful for roofing

Trees and shrubs that grow on chalky soils (*left*) do not grow on heathland (*right*): the chemical composition of the two soils is quite different

Leaded glass windows are still made by traditional methods

About 20 tonnes of aluminium go into each Concorde

QE2 contains 16 000 tonnes of steel

and for leaded windows. It also has a high density and because of this is used in yacht keels and plumb-lines.

The non-metals, silicon and germanium, conduct electricity, though not nearly so well as metals: for this reason they are known as semiconductors. When made into transistors they are the basis of modern electronics. Chlorine, a reactive gas, which is poisonous in large quantities, is used to kill bacteria in drinking water and swimming pools.

A property of iron, which increases its usefulness, is that it can be mixed with other metals to form **alloys**. An alloy is a mixture of two or more metals.

Alloys have different properties from pure metals and it is possible to make alloys with specified properties. The most widely used iron alloy is steel. 54 000 tonnes of steel were used in constructing the Forth rail bridge. Other major uses are in ships, cars and large buildings. Stainless steels contain the metals chromium and nickel.

Another strong metal, used extensively in alloys, is aluminium. Aluminium has a low density and does not rust. For these reasons it is used in large quantities for aircraft construction.

Nowadays, larger amounts of iron and aluminium are used than any other metal. These two metals are also the most abundant metals in the earth's crust.

4.9 What are the most abundant elements?

IN THE EARTH

Figure 4.10 shows the abundance of elements in the earth's atmosphere and crust to a depth of 40 km. Very few elements occur naturally. Most occur as compounds; that is, combined with other elements.

Half of the weight of the earth's crust is made up of oxygen. The oxygen forms 23 % (by weight) of the air; it is also combined with hydrogen in water (89 %), but most of the oxygen is combined with other elements in rocks and soils. Sand contains 53 % oxygen and 47 % silicon. This accounts for the large percentage of silicon in the earth's crust. Clays also contain silicon and oxygen together with aluminium, which is the third most abundant element. Chalk and limestone (calcium

Figure 4.10

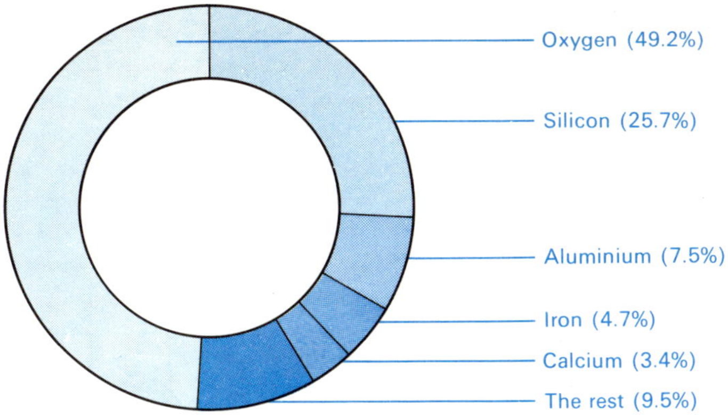

Sandstone

Limestone

A kidney-shaped piece of haematite

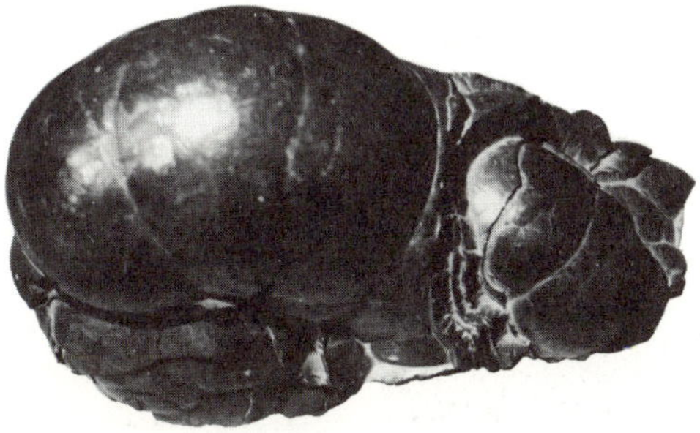

carbonate) contain 40% calcium, 12% carbon and 48% oxygen. This accounts for most of the calcium in the earth's crust. Most of the iron is combined with oxygen in haematite (rust-coloured iron oxide).

Which five elements make up 90% of the earth's crust? The other 100 or so elements make up only 10% of the earth's crust. Which elements do you think form a large portion of this remaining 10%?

IN THE HUMAN BODY

Figure 4.11 shows the abundance of different elements in the human body. Oxygen, carbon and hydrogen are the three most abundant elements in the body. They are the main elements in fats, proteins and carbohydrates which make up most of our body tissues. Water, which makes up about 75% of body weight, accounts for more hydrogen and oxygen. Nitrogen occurs in proteins, and calcium is combined with phosphorus and oxygen as calcium phosphate in bones.

Figure 4.11

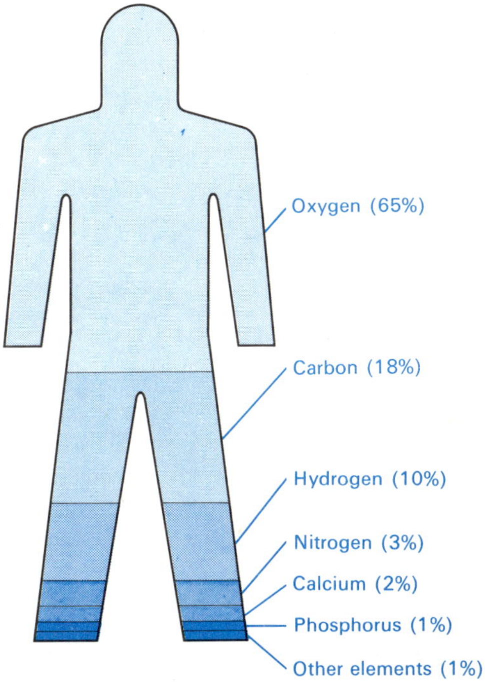

Summary

1 An element is a substance which cannot be broken down into simpler substances by chemical means.

2 Electrolysis (electrical decomposition) is the decomposition of a substance by an electric current.

3 All matter in the universe is built up from elements. Of the 105 known elements, 90 or so exist in the earth's crust or atmosphere. 90% of the earth's crust is composed of 5 elements (oxygen, silicon, aluminium, iron and calcium).

4 Elements are classified as metals and non-metals. Metals usually have a high melting point, a high boiling point, a high density, good conductivity and a shiny surface. Non-metals usually have a low melting point, a low boiling point, a low density, poor conductivity and a dull surface.

5 The uses of elements depend upon:
(a) their properties
(b) the ease with which they can be obtained.

6 Elements can combine together to form compounds. A compound is a substance which contains two or more elements chemically combined together. When elements combine together a chemical reaction takes place and the reaction is usually exothermic. An exothermic reaction is one in which heat is given out.

7 In a mixture no new compound is formed and no chemical reaction takes place.

8 A pure compound always contains the same elements combined together in the same proportions by weight (law of constant composition). A mixture can have any composition.

9 Compounds can be produced from their component elements by combining them together under suitable conditions. The elements can only be regenerated from the compounds by chemical reactions.

Study questions

1 The results in Figure 4.12 were obtained when three solids, *X*, *Y* and *Z*, were heated separately in crucibles.
(a) Make a table of the results in Figure 4.12 like this:

	X	*Y*	*Z*
Weight of solid before heating			
Weight of residue after heating			

(b) Which of the solids may have decomposed on heating?
(c) Which of the solids may have undergone synthesis on heating?
(d) Which of the solids may be the element, magnesium?
(e) Which of the solids may be the element, carbon?
(f) Which of the solids may be the compound, red mercury oxide?
(g) Which of the solids may be the compound, sodium chloride?
(h) Could *Z* be an element? Explain your answer.

2 Synthesis is building up substances but decomposition is breaking down substances.
(a) Is it possible to decompose an element?
(b) Is it possible to decompose a compound?
(c) Which of the following can be decomposed?

Potassium chloride   Silicon
Sugar   Zinc
Brass (an alloy of copper and zinc)

(d) During decomposition two or more simpler substances are always formed. Could all the simpler substances produced be elements? Explain, with an example. Could all the simpler substances produced be compounds? Explain, with an example.
(e) Why must the product of synthesis always be a compound?
(f) Which of the five substances in part (c) could be formed by synthesis?

3 You have been given a red substance, *A*, and a yellow substance, *B*, which are known to react together when heated.
(a) Is it easier to form a 50 : 50 mixture of *A* and *B*, or the compound from *A* and *B*? Explain.

Figure 4.12

	X	Y	Z
Weight of crucible alone	16.22 g	13.47 g	13.41 g
Weight of crucible+solid before heating	22.01 g	13.77 g	22.45 g
Weight of crucible+residue after heating	19.46 g	13.82 g	22.45 g

Figure 4.13

To investigate the percentages of magnesium and oxygen in magnesium oxide	
Readings	
Weight of crucible+lid alone	8.265 g
Weight of crucible+lid+magnesium	8.625 g
Weight of crucible+lid+magnesium after 5 min heating	8·814 g
Weight of crucible+lid+magnesium after 7 min heating	8.865 g
Weight of crucible+lid+magnesium after 9 min heating	8.865 g

Procedure
The crucible and lid were weighed together and then re-weighed with a small piece of magnesium. The magnesium was heated strongly. From time to time during the heating, the bunsen was removed and the lid of the crucible was raised. After 5 min heating the crucible was allowed to cool and was re-weighed. The process of heating, cooling and re-weighing was repeated until the contents of the crucible had been heated to constant weight.

(b) What colour do you think the 50 : 50 mixture will be?
(c) Is it possible to predict the colour of the compound?
(d) Is it possible to predict the percentage of *A* in the compound?
(e) What test would you carry out on each substance to decide whether it is a pure compound or a mixture?

4 (a) Mercury is the only liquid metal at room temperature. It has a high boiling point, a high density and is a good conductor of heat and electricity. What uses does mercury have on account of these properties?
(b) Platinum is a hard unreactive metal with a high melting point. What uses does platinum have on account of these properties?

5 You are given a chunk of solid material. Describe carefully the experiments you would carry out to investigate whether it is an element or a compound:
If it is an element, how would you determine whether it is a metal or a non-metal?

6 The properties of a compound are not the same as those of its component elements. Illustrate this statement by drawing up a table summarizing:
(a) The properties of mercury, oxygen and red mercury oxide.
(b) The properties of iron, sulphur and iron sulphide.

7 Figure 4.13 is an extract from a student's practical book.
(a) Write an equation for the reaction which takes place when magnesium is heated.
(b) What is the weight of magnesium which reacts?
(c) What is the weight of magnesium oxide produced?
(d) What weight of oxygen reacts with the magnesium?
(e) What is the percentage of magnesium in the magnesium oxide?
(f) What is the percentage of oxygen in the magnesium oxide?
(g) Why is it necessary to remove the lid of the crucible from time to time during the heating?
(h) Why must the crucible be cool before it

Figure 4.14

Element	Density at room temperature in g cm^{-3}	Boiling point in °C	Electrical conductivity
A	3.12	58	Poor
B	8.65	765	Good
C	19.3	2970	Good
D	3.4×10^{-3}	−152	Poor
E	0.53	1330	Good
F	2.07	445	Poor

Figure 4.15

	Weight of copper oxide taken in g	Weight of copper formed in g
Experiment 1	7.95	6.35
Experiment 2	9.54	7.62

is re-weighed?

(i) What is meant by 'heating to constant weight'?

(j) Why is it necessary to heat to constant weight?

8 Some properties of six elements labelled *A* to *F* are given in Figure 4.14.

(a) Which of the six elements are metals?

(b) Which element will float on water as a solid at room temperature?

(c) Which element is a gas at room temperature?

(d) Which non-metal may be a solid at 60°C?

(e) Which metal is probably a liquid at 2000°C?

(f) What is the boiling point of the least dense metal?

(g) The elements in the list are cadmium, gold, lithium, bromine, krypton and sulphur. Which element is which?

9 (a) Which is the most abundant element in the earth's crust?

(b) Why do you think this element is so abundant? What substances found in the earth's crust contain this element?

(c) Which metals are likely to be next most abundant after aluminium, iron and calcium? Give reasons for your choice.

(d) Name two elements which exist uncombined in the atmosphere.

(e) Name two elements which exist uncombined in the earth's crust.

(f) Why is iron used more widely than aluminium, even though the earth's crust contains twice as much aluminium as iron?

10 When anhydrous copper sulphate is added to water, blue copper sulphate is produced. Give three pieces of evidence to show that the product is a compound and not a mixture.

This reaction is exothermic. Mention one exothermic process which is important in the home.

11 When 1 g of solid *A* was heated, $1\frac{1}{2}$ g of solid *B* were formed. No change occurred on further heating.

Which of the following statements are correct?

(a) Synthesis may have occurred.

(b) *A* may be an element.

(c) *A* may be a compound.

(d) *B* may be a mixture.

(e) Decomposition may have occurred.
(f) *A* may be a mixture.
(g) *B* may be an element.
(h) *B* may be a compound.

12 When a lighted taper is placed in an inverted test tube containing hydrogen the hydrogen explodes with a 'pop'.
(a) What gas in the air does hydrogen react with in the 'pop' test?
(b) What will the product of the reaction be?
(c) Write a word equation for the reaction?
(d) Describe carefully the experiments you would do to show that water is a compound of hydrogen and oxygen only.
(e) Which of the following statements are true?
Water is classed as a compound rather than an element because it:
(i) is not easily decomposed.
(ii) has a fixed melting point and boiling point at atmospheric pressure.
(iii) is formed when hydrogen reacts with oxygen.
(iv) is a pure substance.

13 Figure 4.15 relates to experiments which were carried out to test the law of constant composition. Two samples of copper oxide were converted to copper .
(a) Calculate the weight of oxygen in each sample of copper oxide.
(b) Calculate the percentage of oxygen in each sample of copper oxide.
(c) Calculate the percentage of copper in each sample of copper oxide.
(d) What does the law of constant composition state?
(e) Do these results agree with this law?

14 Give one important use of each of the following metals and one reason why the metal is so suitable for that use.

Copper	Zinc	Tin
Calcium	Aluminium	Silver

15 What elements do the following contain?

Red mercury oxide	Sugar
Octane (in petrol)	Bread
Marble (calcium carbonate)	Water
Copper sulphate crystals	Alcohol
Anhydrous copper sulphate	A penny
Sodium chloride	Cheese
Polythene	Wood

Testing conductance

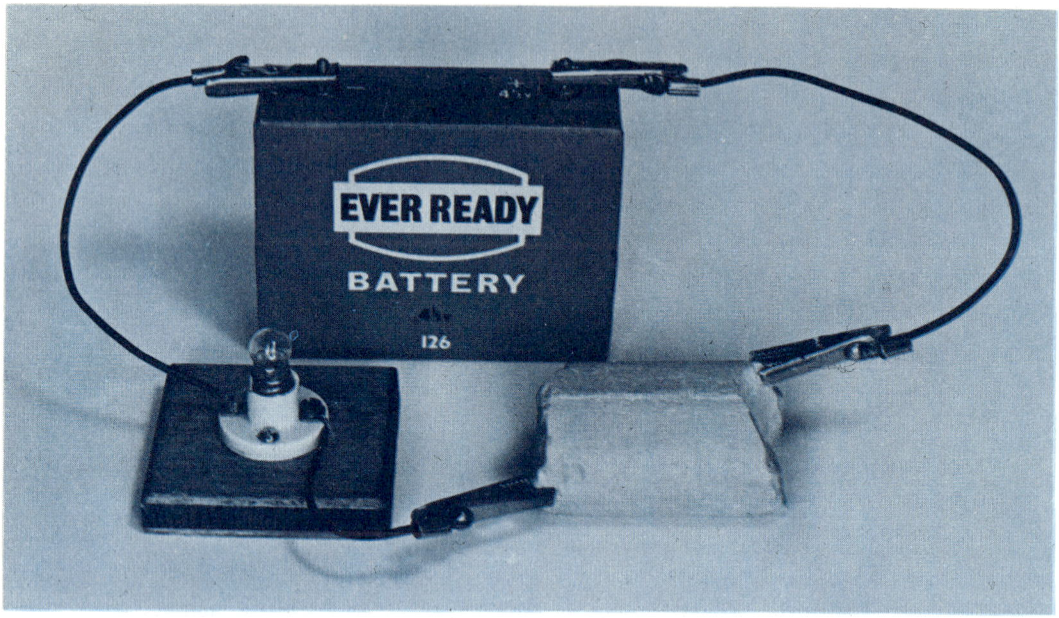

5 Action of electricity on materials

5.1 Introduction

In the last chapter we used the apparatus in Figure 4.4 to determine which elements conducted electricity.

Try to answer these questions

1 What happens in the circuit of Figure 4.4 if the element conducts? What happens if the element does not conduct?

2 What general summary can be made about the conduction of electricity by elements?

3 Do the elements which conduct change in weight or appearance during conduction?

5.2 Do solid compounds conduct electricity?

The same circuit as Figure 4.4 can be used to see if solid compounds conduct electricity. The bulb never lights, showing that solid compounds are all very poor conductors. In fact, they are such poor conductors that they can be described as non-conductors or insulators.

Thus, the only solids which conduct electricity are metals and graphite.
Metals do not react when they conduct and no new substance forms; nor is there any change in weight or appearance. However, if the metal is in the form of a very thin wire and the current is large, it is possible to make the wire red hot. In some cases, the wire may get so hot that it reacts with oxygen in the air.

5.3 The uses of conductors and insulators

Copper, nickel, iron, aluminium and silver, which are good electrical conductors and which are easily made into wire, are used for power lines, cables, fuses and wiring in electrical machinery. Porcelain, and plastics such as polythene, cellulose acetate and polyvinyl chloride (PVC), which are very poor conductors, are often used to insulate wires and cables.

A number of substances with special and complex conducting properties are called semiconductors. Silicon containing a trace of indium is a semiconductor, an important part of any transistor.

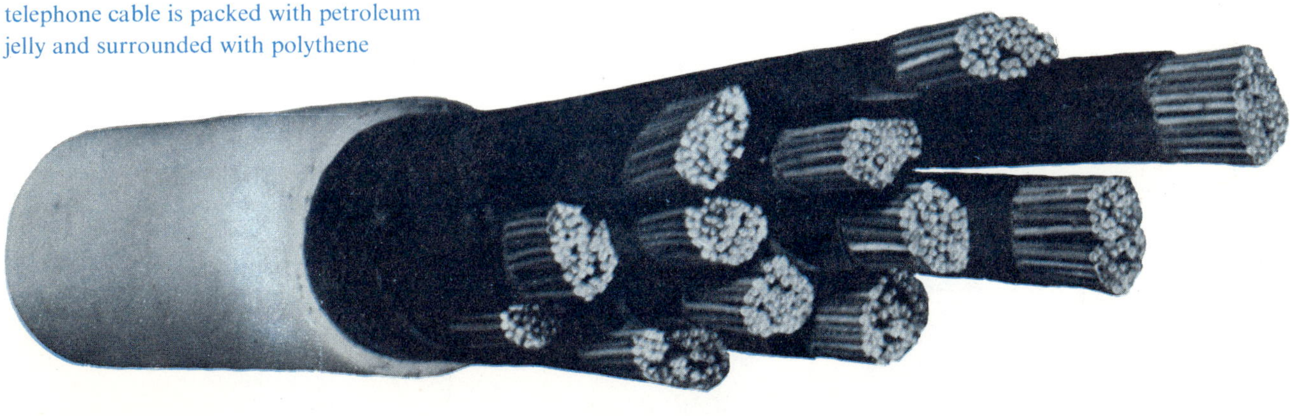

Each bundle of aluminium wires in this telephone cable is packed with petroleum jelly and surrounded with polythene

5.4 Do liquids conduct electricity?

We know that water containing a little sulphuric acid will conduct electricity (Section 4.2, page 41). Unlike metals, water undergoes a chemical reaction when it conducts electricity. The water is decomposed by electrical energy into hydrogen and oxygen.

Electricity, like heat, is a form of energy, and we can use it to cause changes in materials. We use it in electric kettles to boil water and in electric stoves to cook food. The decomposition of a substance, such as water containing a little sulphuric acid, by electricity is called **electrolysis.** We say that the substance decomposed is an electrolyte, and we say it has been electrolysed. Figure 5.1 shows a modified form of the apparatus in Figure 4.4 (page 43) which can be used to test the conductivity of pure liquids, molten solids and solutions.

The terminals through which the current enters and leaves the electrolyte are called **electrodes.** The electrode connected to the positive terminal of the battery is positive itself and is called the **anode.** The electrode connected to the negative terminal of the battery is negative itself and is called the **cathode.** Figure 5.2 shows the results of conduction tests on various liquids.

Try to answer these questions

Look at the results in Figure 5.2.

1 Which liquid elements conduct electricity?

2 Which liquid elements do not conduct electricity?

3 Which liquid compounds conduct electricity?

4 Which liquid compounds do not conduct electricity?

5 Do all liquid compounds which conduct contain a metal?

6 Do all aqueous solutions which conduct contain a metal?

The important points to be made from these results are:

1 The liquid metals, mercury and molten zinc, both conduct electricity, but neither of the liquid non-metals, sulphur and bromine, does. In general, molten metals are conductors but liquid non-metals are non-conductors. Molten metals are not decomposed during conduction.

2 Pure water is a very poor conductor of electricity and can be classified as a non-conductor.

3 Compounds which contain both metals and non-metals (called metal/non-metal compounds) conduct electricity in the liquid state. Notice that water is not essential for conduction of electricity by these compounds. Examples of such compounds are lead bromide, potassium iodide and sodium chloride. These compounds are changed into other substances during conduction. Later in this chapter we shall investigate the products formed during conduction.

4 Compounds which contain only non-metals (called non-metal/non-metal compounds) do not conduct electricity in the liquid state. Examples of such compounds are water, ethanol, benzene, wax, naphthalene and sugar. Benzene, wax and naphthalene contain the elements carbon and hydrogen only; ethanol and sugar contain oxygen as well.

5 Compounds which contain both metals and non-metals (and acids) conduct electricity in aqueous solution. Aqueous solutions of lime water (calcium hydroxide), potassium iodide, copper sulphate and sodium chloride are examples of aqueous solutions of metal/non-metal compounds which conduct electricity. Aqueous solutions of sulphuric acid and acetic acid also conduct electricity although neither contains a metal.

The varying brightness of the bulb shows that some of these aqueous solutions conduct better than others. For example, aqueous solutions of copper sulphate and sodium

Figure 5.1

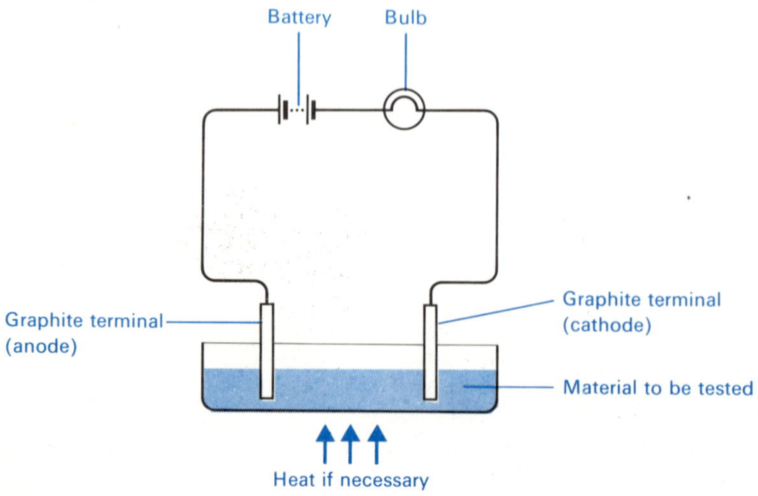

chloride conduct better than aqueous solutions of calcium hydroxide (lime water) and acetic acid (vinegar).

New substances are produced at the electrodes when aqueous solutions of metal/non-metal compounds are electrolysed.

6 Compounds which contain only non-metals (except acids) do not conduct electricity in aqueous solution. Aqueous ethanol and aqueous sugar are examples of aqueous non-metal/non-metal compounds which do not conduct electricity. Apart from a few exceptions, acids are the only aqueous non-metal/non-metal compounds to conduct electricity. Thus, vinegar (acetic acid) and sulphuric acid both conduct electricity. Acetic acid contains carbon, hydrogen and oxygen; sulphuric acid contains sulphur, hydrogen and oxygen.

The results of these experiments are summarized in Figure 5.3.

Figure 5.2

Pure liquids	**Does the bulb light, showing that the liquid conducts?**
Distilled water	No
Ethanol	No
Benzene	No
Mercury	Yes
Molten zinc	Yes
Molten wax	No
Molten naphthalene	No
Molten sulphur	No
Bromine	No
Molten lead bromide	Yes
Molten potassium iodide	Yes
Molten sodium chloride	Yes

Solutions	**Does the bulb light, showing that the solution conducts?**
Aqueous ethanol	No
Aqueous sugar	No
Aqueous sulphuric acid	Yes
Aqueous vinegar (acetic acid)	Yes (poor)
Aqueous lime water (calcium hydroxide)	Yes (poor)
Aqueous potassium iodide	Yes
Aqueous copper sulphate	Yes
Aqueous sodium chloride	Yes

Figure 5.3: Do they conduct electricity?

	Elements		**Compounds**	
	Metals and graphite	**Non-metals except graphite**	**Metal / non-metal**	**Non-metal / non-metal**
Solid	Yes	No	No	No
Liquid	Yes	No	Yes	No
Aqueous solution	———	———	Yes (and acids)	No (except acids)

Try to answer this question
Copy out Figure 5.4 and fill in the empty spaces. You will find it helpful to refer to Figures 4.7 and 5.3.

Figure 5.4

Substance	Conduction in the solid state	Conduction in the liquid state	Conduction in aqueous solution	Are new substances formed during conduction in the liquid state?
Iron		Yes		
Phosphorus			No (does not dissolve)	
Diamond (carbon)	No			
Carbon tetrachloride (tetrachloromethane)				No
Zinc bromide			Yes	

Figure 5.5

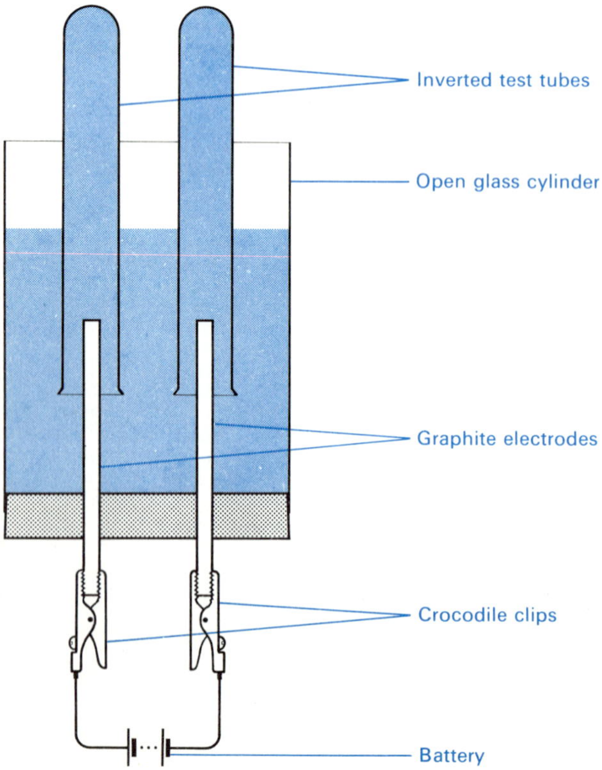

5.5 What substances are formed at the electrodes?

Acids and metal/non-metal compounds that conduct electricity are changed into new substances during conduction. For example, molten lead bromide produces brown fumes of bromine at the anode and lead at the cathode. When the molten lead bromide has cooled down, you will find a bead of lead in the solidified electrolyte beneath the cathode. When electricity is passed through molten potassium iodide, purple fumes of iodine vapour come off at the anode and potassium forms at the cathode. With copper sulphate solution, a pink deposit of copper appears at the carbon cathode, and bubbles of a colourless gas stream away from the carbon anode.

Several other aqueous solutions produce bubbles of gas at the electrodes. The apparatus in Figure 5.1 is unsuitable for collecting and testing these gases. Figure 5.5 shows a modified version for examining the products of electrolysis of aqueous solutions. Bubbles produced at the electrodes rise through the solution and collect in the inverted test tubes. Colourless gases, such as hydrogen and oxygen can be identified by the usual tests.

Figure 5.6 lists the products formed at each of the electrodes when various liquids and aqueous solutions are electrolysed. Of course, water may take part in the electrolysis of aqueous solutions.

Try to answer these questions
Look at Figure 5.6.
1 Which elements are produced at the anode?
2 Which elements are produced at the cathode?

When molten or aqueous metal/non-metal compounds conduct electricity:
1 Metals or hydrogen are formed at the cathode.
2 Non-metals (except hydrogen) are formed at the anode.

In every case, an element was produced at each electrode. Thus, electrical decomposition of the metal/non-metal compound occurs. This is very different from the conduction of electricity by metal elements which are not decomposed during conduction. The first three electrolyses in Figure 5.6 are summarized below in word equations:

$$\text{lead bromide}_{(l)} \xrightarrow[\textit{energy}]{\textit{electrical}} \text{lead}_{(l)} + \text{bromine}_{(g)}$$

$$\text{potassium iodide}_{(l)} \xrightarrow[\textit{energy}]{\textit{electrical}} \text{potassium}_{(l)} + \text{iodine}_{(g)}$$

$$\text{sodium chloride}_{(l)} \xrightarrow[\textit{energy}]{\textit{electrical}} \text{sodium}_{(l)} + \text{chlorine}_{(g)}$$

The fact that electrolytes can be decomposed by electrical energy suggests that the 'glue' holding elements together in metal/non-metal compounds is electrical in some way.

Figure 5.6

Substance tested	Product at anode	Product at cathode
Molten lead bromide	Brown fumes of bromine	Lead
Molten potassium iodide	Purple fumes of iodine	Potassium
Molten sodium chloride	Pale green chlorine gas	Sodium
Aqueous potassium iodide	Iodine which colours the solution brown	Hydrogen
Aqueous copper sulphate	Oxygen	Copper (deposited on the cathode)
Aqueous sodium nitrate	Oxygen	Hydrogen
Aqueous zinc bromide	Bromine which colours the solution brown	Zinc (deposited on the cathode)
Aqueous copper chloride	Chlorine	Copper (deposited on the cathode)

5.6 Classifying elements as metals and non-metals

The action of electricity on elements and compounds has revealed clear differences between metals and non-metals and between their compounds.
1 Metal elements are good conductors of electricity, but non-metals (except graphite) are very poor conductors.
2 Molten and aqueous metal/non-metal compounds conduct electricity but molten and aqueous non-metal/non-metal compounds (except aqueous acids) do not conduct electricity.
3 When electrolysis occurs, metals are formed at the cathode and non-metals (except hydrogen) are formed at the anode.

These differences provide still more evidence that the classification of elements as metals and non-metals is a satisfactory and useful one.

5.7 The industrial applications of electrolysis

The results of our experiments illustrate two important industrial uses of electrolysis:
1 Extracting elements from their compounds.
2 Electroplating, depositing a thin layer of metal on an object by passing an electric current.

EXTRACTING ELEMENTS BY ELECTROLYSIS

Sodium, magnesium, calcium, aluminium, chlorine, oxygen and hydrogen are extracted from their compounds by electrolysis. The first three are manufactured from their molten chlorides, the electrolysis yielding chlorine, a valuable by-product, at the anode. Aluminium is obtained from aluminium oxide.

Large quantities of electricity are needed for these electrolyses, which is why extraction plants are situated near supplies of cheap electricity. (It takes about 20 kWh of electricity to produce 1 kg of aluminium from the oxide.) Most of Britain's aluminium plants are in Scotland; they generate their own hydro-electricity.

Hydrogen and oxygen can be produced as gases by the electrolysis of dilute sulphuric acid, but on a large scale this is only done if cheap electricity is available. Some compounds are manufactured by electrolysis. The electrolysis of concentrated brine (sodium chloride solution) yields chlorine at the anode and hydrogen and sodium hydroxide (caustic soda) at the cathode. The sodium hydroxide and chlorine are used to manufacture sodium hypochlorite (used in bleaches) and sodium chlorate (used in weed killers). Sodium hydroxide is also used to make soap and sodium carbonate (known as washing soda).

ELECTROPLATING

The metals used most commonly for electroplating are copper, chromium, silver and tin. Although it is simple to deposit a metal during electrolysis, the conditions must be carefully adjusted so that the metal deposit will adhere (stick) to the object to be plated. Chromium will not adhere to iron or copper. Thus, iron or copper articles to be chromium plated are first plated with nickel which adheres strongly. The nickel coating is then plated with chromium.

In order to obtain a good coating of metal:

1 The object to be plated must be clean and free of grease.

2 The object to be plated must be rotated to give an even coating.

3 The electric current must not be too large or the 'coating' will form too rapidly and will flake off.

4 The temperature and concentration of the electrolyte must be carefully controlled, otherwise the 'coating' will be deposited either too rapidly or too slowly.

The metal coating is deposited on the cathode and so the object to be coated must be made the cathode during electrolysis. The electrolyte must contain a compound of the metal which forms the coating. Figure 5.7 shows a nickel alloy knife being electroplated with silver.

Try to answer these questions

Steel bumpers on cars are electroplated first with nickel and then with chromium.

1 Why is the steel plated with a layer of nickel before chromium?

2 Give two reasons for electroplating steel.

3 Cutlery made of nickel alloy electroplated with silver is stamped EPNS. What does EPNS stand for?

4 Articles of jewellery (e.g., bangles) are often electroplated with gold, silver or copper. Why?

Sodium hypochlorite is the active ingredient in many 'sterilizing' liquids

Sodium chlorate kills weeds

Summary

1 Metals and graphite conduct electricity in the solid and liquid states. Non-metals (except graphite) do not conduct electricity in the solid and liquid states. The metals and graphite are not decomposed during conduction.

2 Compounds do not conduct electricity in the solid state.

3 (a) Metal/non-metal compounds conduct electricity when molten or in aqueous solution.
(b) Non-metal/non-metal compounds do not conduct electricity when molten or in aqueous solution. Acids are an exception to this general rule since they conduct electricity in aqueous solution.

4 Compounds that conduct electricity when molten or in aqueous solution are called electrolytes. They are decomposed (electrolysed) during the process.

5 When electrolysis occurs:
(a) a metal or hydrogen is produced at the cathode.
(b) a non-metal (except hydrogen) is produced at the anode.

6 Electrolysis is used in industry to extract elements from their compounds and to electroplate objects.

Study questions

1 Water containing a little dilute sulphuric acid was decomposed when electrolysed in the apparatus shown in Figure 5.5.
(a) What products are formed when water is decomposed? Write a word equation to summarize the results.
(b) Explain clearly what is meant by the word 'decompose'.
(c) Where does the energy needed to decompose the water come from?
(d) What substance is produced at the anode?
(e) What substance is produced at the cathode?
(f) How would you test to see if the products you suggest are formed?
(g) What would happen if you changed round the connections from the battery to the electrodes?
(h) In some countries hydrogen is obtained from solutions containing water by electrolysis. What do you think is the most expensive aspect of these electrolyses? Which countries are most likely to use electrolytic processes to obtain hydrogen?

2 Describe carefully two experiments you would carry out to determine whether a chunk of solid is a metal or a non-metal.
(a) Name four elements which conduct

Figure 5.7

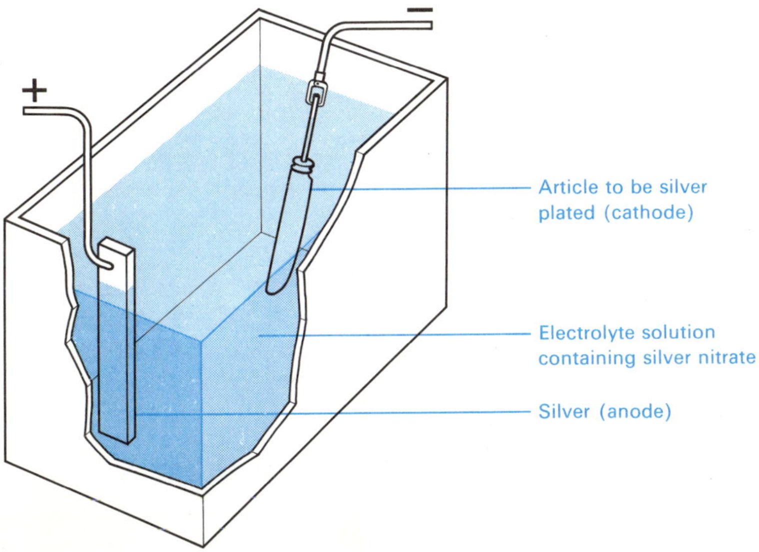

This electroplated oak leaf will be made into a piece of jewellery

electricity when solid.
(b) Name four elements which conduct electricity when liquid.
(c) Name four elements which do not conduct electricity when solid or liquid.
(d) Name four compounds which conduct electricity when liquid.
(e) Name four compounds which do not conduct electricity when liquid.

3 Calcium, carbon disulphide, copper sulphate solution, lead, carbon, water, methane (the main constituent of North Sea gas), phosphorus, dilute sulphuric acid. From the above list name:
(a) two metals.
(b) two non-metals.
(c) two pure liquids at 20°C.
(d) two compounds that are gases at 110°C.
(e) two elements which conduct electricity.
(f) two electrolytes.
(g) three compounds that are non-electrolytes.

4 (a) Why is an object to be electroplated made the cathode of an electrolytic cell?
(b) Suppose you want to silver-plate a spoon. What substance would you choose for:
(i) the anode.
(ii) the electrolyte?
(c) Give two reasons why articles are electroplated.
(d) What precautions are taken to obtain a good metal coating?
(e) What happens if the metal is deposited too quickly?

5 When substance, *X*, was electrolysed two different pure substances were obtained.
(a) Could *X* be an element?
(b) Could *X* be a pure compound?
(c) Could *X* be a mixture?
(d) Could *X* be a solution?
Give an explanation of your answers in (a)–(d). Give an example to illustrate each of the suggestions you think is true.

6 Copy out and complete Figure 5.8.

7 Copy out and complete Figure 5.9.

8 Substances *A*–*E* in Figure 5.10 are mercury, sulphur, potassium chloride, sugar (a compound of carbon, hydrogen and oxygen) and hydrogen chloride. Which letter corresponds to which substance?

9 Which of the following answers are true?
(a) When an aqueous solution conducts

Figure 5.8

Molten substance electrolysed	Product at anode	Product at cathode
Copper chloride		
	Bromine	Potassium
Aluminium oxide		
	Iodine	Sodium
Lead bromide		

Figure 5.9

Substance	Conduction in the solid state	Conduction in the liquid state	Conduction in aqueous solution	Are new substances formed during conduction?
Sugar				
Sodium chloride				
Hydrogen chloride				
Carbon tetrachloride (tetrachloromethane)			Insoluble in water	
Carbon			Insoluble in water	
Magnesium			Insoluble in water	

electricity you will always observe:
(i) a chemical change.
(ii) gas produced from at least one electrode.
(iii) new substances produced at the electrodes.
(iv) electroplating.
(v) a metal produced at the cathode.
(b) When electrolysis occurs you may observe, at the negative electrode, the formation of:
(i) a metal.
(ii) a non-metal.
(iii) oxygen.
(iv) hydrogen.
(v) a liquid.

10 A non-conductor, *X*, melts at a high temperature to form a liquid which conducts electricity. Which of the following could *X* be?

Calcium chloride	Polythene
Starch	Copper
Bronze	

11 You are required to plate a graphite rod with copper. Draw a labelled diagram of the apparatus you would use to do this.

12 The main ore from which aluminium is extracted is bauxite (impure aluminium oxide). The bauxite is first purified to obtain aluminium oxide. The pure aluminium oxide is dissolved in another molten electrolyte (cryolite) and this solution is then electrolysed between carbon electrodes. The process is only economic if cheap electricity is available.
(a) Aluminium oxide is decomposed during the electrolysis. What are the products at the anode and at the cathode?
(b) The carbon anodes are gradually burnt away. What does the carbon of the anode react with?
(c) Why does the carbon of the cathode not get burnt away?
(d) Why is most of Britain's aluminium produced in the Scottish Highlands?
(e) What are the main advantages in using aluminium rather than iron in the construction industry?

13 Figure 5.11 shows the apparatus used to purify copper by electrolysis. *A* and *B* are the terminals of the battery and pure copper is slowly deposited on *C*.
(a) Is *C* the anode or the cathode?
(b) Is *B* connected to the positive or negative side of the battery?
(c) What material is used for *C*?
(d) Name the electrolyte, *D*.

Figure 5.10

Substance	M. pt. in °C	B. pt. in °C	Electrical conductivity of aqueous solution
A	−114	−85	Good
B	170	Decomposes	None
C	−38	357	Substance insoluble
D	770	1407	Good
E	119	445	Substance insoluble

Figure 5.11

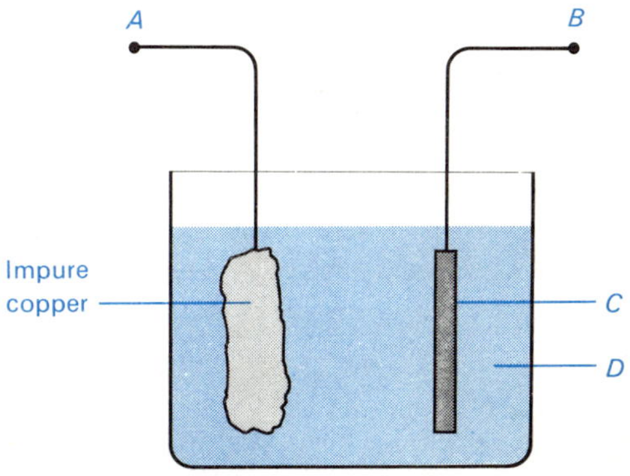

'Breathe in that fine air'

6 Burning, breathing and rusting

6.1 Introduction

What would happen if all the air was suddenly removed from the Earth? Would coal and gas fires continue to burn? Would we continue to live?

Our very existence depends on a good supply of air, yet of all the materials in the Universe, air is one that we think very little about – until something goes wrong with it or until we are deprived of it. In this chapter and the next we will see just how important the air is in everyday life; in burning, breathing and rusting.

An aboriginal lights a fire with a drill

6.2 Burning

Burning was the first chemical process used by primitive man. He used it to provide himself with warmth, to cook his food by, and later to produce new materials, such as metals and clay pots.

It is just as essential to our lives as it was thousands of years ago to the lives of our ancestors. Burning fuel keeps us warm, drives our motor cars and produces steam to drive the turbines which generate most of our electricity.

Try to answer these questions

When copper or magnesium are heated in air they gain weight.

1 What are the names of the substances produced?

2 Where has the increase in weight come from?

3 Is anything removed from the air when copper or magnesium are heated?

4 Write a word equation for the reaction that occurs when magnesium is heated in air.

Today we use rather more sophisticated methods of cooking

Other substances, such as red mercury oxide, copper carbonate and blue copper sulphate lose weight on heating. These substances decompose on heating to form two or more products. One of these products is a gas which escapes into the air.

We know that many reactions which happen when substances are heated are either synthesis or decomposition. But what about burning?

What reaction does burning involve? Do burning substances combine with oxygen from the air (like heated copper or magnesium), or do burning substances decompose (like blue copper sulphate)? The two experiments which follow will give us a clue.

Figure 6.1

EXPERIMENT 1. What happens when a candle burns?

Burn a small candle on a watch glass under a large beaker (Figure 6.1). Block up the spout of the beaker with Plasticine.

(a) The candle soon goes out, unless more air is admitted into the beaker. A lighted splint goes out if it is placed in the remaining gas.

(b) A mist forms on the inside of the beaker. The mist is due to tiny droplets of a colourless liquid, which turns white anhydrous copper sulphate blue.

(c) The gas remaining in the beaker turns lime water milky.

Try to answer these questions

1 Why does the candle go out?

2 Which gas has been removed from the air during burning?

3 Which product forms the mist on the inside of the beaker?

4 Which product turns the lime water milky?

The reaction which takes place can be summarized as

$$\text{candle}_{(s)} + \text{oxygen}_{(g)} \longrightarrow \text{water}_{(g)} + \text{carbon dioxide}_{(g)}$$
(*hydrogen oxide*)

The candle uses up oxygen from the air to form water and carbon dioxide. This is similar to the reaction involved when copper is heated in air.

$$\text{copper}_{(s)} + \text{oxygen}_{(g)} \longrightarrow \text{black copper oxide}_{(s)}$$

EXPERIMENT 2. What changes in weight occur when a candle burns?

Copper gains weight when it is heated in air. The black copper oxide that forms weighs more than the original copper.

How does a candle change in weight when it burns? It loses weight! Why does it lose weight? Because the water and carbon dioxide produced are gases and these escape into the air.

What happens if we collect the products of combustion (water and carbon dioxide) and then investigate the weight changes? It is possible to weigh the water and carbon dioxide by absorbing them, as soon as they are produced, in a substance called soda lime. The apparatus designed to do this is shown in Figure 6.2.

First, weigh the entire apparatus (candle + watch glass + filter funnel + U tube + delivery tubing).

Then, draw a slow stream of air through the apparatus with the pump, and light the candle. Let the candle burn for about 5 min, disconnect the pump and weigh the entire apparatus again when it is cool.

Figure 6.2 also shows a typical result.

Figure 6.2

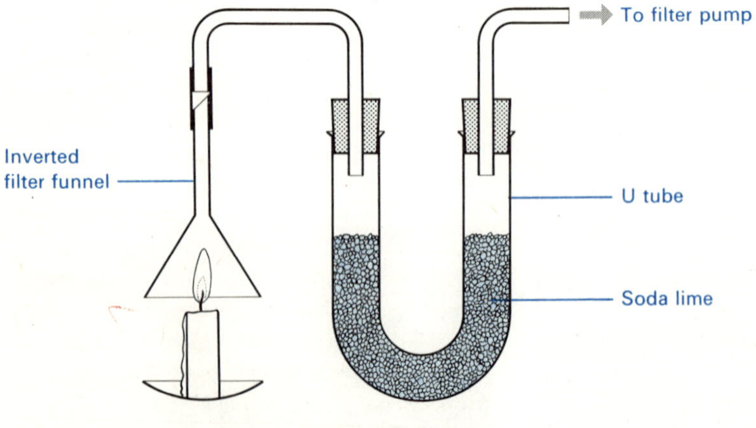

Weight of entire apparatus before burning = 129.81 g
Weight of entire apparatus after burning = 130.30 g

Try to answer these questions

1 Has the entire apparatus increased or decreased in weight on burning?

2 Will the candle by itself increase or decrease in weight on burning?

3 How do you explain the change in weight of the entire apparatus?

During this experiment the candle by itself loses weight, but the soda lime increases in weight by a greater amount. Thus, the entire apparatus increases in weight during burning.

This means that the products of combustion of candle wax weigh more than the candle wax used up. The increase in weight is due to the candle wax burning in air to form carbon dioxide and water (hydrogen oxide):

$$\text{candle wax}_{(s)} + \text{oxygen}_{(g)} \longrightarrow \text{carbon dioxide}_{(g)} + \text{water}_{(g)}$$

This is further evidence that substances which burn combine with oxygen in the air just like heated metals.

Industry depends on the supply of large quantities of fuel

The products become incandescent

6.3 Why is burning important?

Large quantities of coal, oil and gas are burnt each year in our homes for cooking and heating. Aircraft, space capsules, ships and motor vehicles are driven by the energy produced when fuels burn.

The so-called 'heavy' industries (such as iron and steel, and plastics and chemicals) depend upon the supply of large quantities of fuel. The most important industrial fuels are coal, oil and natural gas. In 1970, industry burnt 160 million tonnes of coal, 90 million tonnes of oil and 20 million tonnes of gas.

Oil and gas are imported into Britain, but during the last few years oil companies have found both of them several hundred metres beneath the North Sea. What effect will this source of cheap fuel have on our lives?

North Sea gas is already beginning to take the place of coal in our homes and in industry. The amount of North Sea gas already found can provide about five times the country's present needs. One quarter of the nation's gas already comes from the North Sea and by 1976, 90% of the country will be using it.

These fuels are composed largely of compounds which contain the elements carbon and hydrogen. During burning these elements combine with oxygen in the air to form carbon dioxide and water. We can summarize this as:

$$\text{fuel} + \text{oxygen} \longrightarrow \text{carbon dioxide} + \text{water} + \text{heat energy}$$

Any substance which burns in air to produce heat energy is a **fuel.**

FLAMES, EXPLOSIONS AND SMOKE

Reactions which produce heat energy are exothermic. Often, so much heat is produced that the products become incandescent (i.e., light energy is produced) forming a flame. In a bunsen burner flame the gases react with oxygen in the air to produce the flame.

Some mixtures of gas, such as hydrogen and oxygen, react so vigorously that enough heat is produced to cause an explosion. When a test tube of hydrogen is tested using the 'pop' test, the hydrogen reacts with oxygen so fast that burning takes place very rapidly. The heat produced causes the gases to expand rapidly and you hear the 'pop' as these gases are expelled from the test tube. In the same wa

T — I — M — B — E — R

Many fuels produce smoke

air is expelled from your mouth as you say the word 'pop'.

Much louder bangs are produced during the rapid combustion of explosives such as TNT and dynamite.

6.4 Pollution

Some fuels, such as natural gas and anthracite, burn without producing smoke or soot. Many other fuels such as oil and most types of coal usually produce smoke. The smoke is small particles of unburnt carbon.

About 3 million tonnes of smoke are produced every year in the United Kingdom from incompletely burnt fuels. This results in the loss of many tonnes of fuel and the pollution of our atmosphere. In heavy industrial areas, about 200 tonnes of grit and dust would be deposited on each square kilometre per year if some means of reducing air pollution was not used.

Each one of us breathes in about 14 000 litres of air every day. Smoke pollution contributes to the cause of bronchitis and other lung diseases.

In 1956, the Clean Air Act made it an offence to pollute the air with sooty smoke from industrial plants, and Local Authorities were given the power to set up smoke control areas ('smokeless zones') to prevent air pollution from homes. Only smokeless fuels, such as coke and anthracite, can be used in smokeless zones.

Britain's first smokeless zone, about a tenth of a square kilometre in the centre of Coventry, came into being on 1 March 1951; a second, almost half a square kilometre, was started in central Manchester in May 1952.

Exhaust fumes from the engines of cars, lorries and trains contain soot and carbon monoxide (a poisonous gas) due to the incomplete combustion of carbon fuels. Engine developments will make exhaust fumes less dangerous.

But air pollution still costs the country more than 50p per family each week in loss of work through illness and in the cost of cleaning buildings.

Compare the cleaned Chapter House with the rest of Westminster Abbey

Figure 6.3

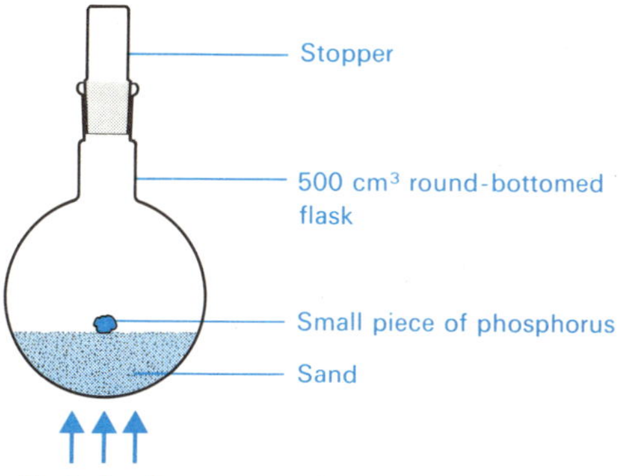

6.5 Changes in weight on burning

Some substances change weight when they burn in oxygen. These substances have either gained something from the air or lost something to it. What happens if the overall weight of the reacting substances (including the air) is taken into account? Weight indicates the amount of matter that is present. The amount of matter is called **mass.**

In order to find out if there is any change in the amount of matter when a chemical change takes place, some phosphorus was heated in a closed stoppered flask (Figure 6.3).

The whole stoppered flask was weighed before and after the phosphorus had reacted with the air inside. Thick white clouds of tiny solid particles are produced. The weight of the flask was the same before and after the reaction. **Don't try this experiment yourself. Phosphorus is very reactive and can burst into flames of its own accord.**

Try to answer these questions

1 Why is it necessary to keep the flask closed during the experiment?

2 What substance in the air has phosphorus combined with?

3 Has the mass changed during the reaction?

4 Do you think it is possible to create mass in a reaction? Is it possible to get something from nothing?

5 Is it possible to destroy mass in a reaction?

During the last century, many chemists carried out experiments similar to this one, to see if the total mass changed after a reaction had taken place. In all these experiments the weights of the reactants and products were equal. There was no observable change in mass (i.e., mass was conserved). Chemists summarized these results in the **law of conservation of mass (matter)**:
The total mass does not change during a chemical reaction; the total mass of reacting substances equals the total mass of the products.

6.6 Chemical laws

A chemical law is a summary of the results of many experiments. Chemical laws are similar in some ways to the laws of society (the 'legal'

In some nuclear explosions, matter is destroyed and vast quantities of energy are produced. The diagonal traces are vapour trails from rockets launched to monitor the fireball

laws). Just as there are a few people in society who disobey the laws of our society so there are sometimes a few reactions which do not obey a particular chemical law.

This is true of the law of conservation of mass. Most chemical reactions obey this law, but in nuclear reactions, such as the explosion of an atomic bomb, matter is destroyed. In spite of this limitation, the law is still useful to chemists for most reactions. If a chemical law is unsatisfactory, it can be changed to take into account its defects, in the same way that a 'legal' law can be changed or repealed if it is unsatisfactory.

6.7 What percentage of the air is oxygen?

What happens when copper is heated in air? Suppose we have 100 cm^3 of air in a closed container: can we find how much air is used up by heating it with copper?

Here is an experiment to measure the amount of air used up. Two syringes are joined through a hard glass silica tube containing copper (Figure 6.4).

Empty one syringe completely by pushing its plunger full in. Fill the other syringe with 100 cm^3 of air. Strongly heat the hard glass tube containing copper. When the copper is red hot move the plungers slowly backwards and forwards during the heating so that all the

Figure 6.4

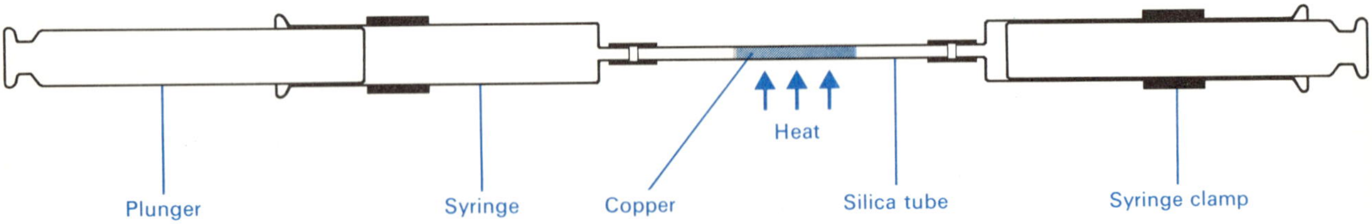

Figure 6.5

Volume of air in syringe before heating	100 cm^3
Volume of air in syringe after heating for 2 min	82 cm^3
Volume of air in syringe after heating for 3 min	79 cm^3
Volume of air in syringe after heating for 4 min	79 cm^3

air comes into close contact with the heated copper. After 2 or 3 min stop heating and cool the tube with a damp cloth. Now measure the volume of air left in the syringe. Repeat the heating for another minute or so, cool as before and continue the heating and cooling until the volume of air which remains is constant.

Figure 6.5 shows some typical results.

Try to answer these questions

1 Has all the oxygen been used up after 2 min?
2 Why is it important to repeat the heating and cooling until the volume of air remaining is constant?
3 How much air did the copper remove?
4 What is the percentage of oxygen in the air?

6.8 Early investigations of burning

The explanation of burning (combustion) was largely the work of two men: Joseph Priestley (1733–1804), a nonconformist clergyman in England who, late in life, became interested in experimental science, and Antoine Lavoisier (1743–1794), the son of a rich French lawyer.

Lavoisier was the first chemist to show that about $\frac{1}{5}$ of air is oxygen. In 1774, Priestley prepared a gas, by heating red mercury oxide, in which substances burnt much more vigorously than they did in air.

The discovery of this gas provided Lavoisier with the vital clue he needed to explain burning. Lavoisier showed that the gas Priestley had made was the same as the $\frac{1}{5}$ part of air which supports burning. He heated mercury gently in a retort containing air for several days. He noticed that the volume of air in the retort and bell jar was reduced by $\frac{1}{5}$.

The reduction in volume was shown by the mercury in the trough rising into the bell jar. At the same time a red solid formed on the surface of the mercury in the retort. Lavoisier showed that substances would not burn in the remaining gas, and small animals such as mice when placed in it very soon died. The remaining $\frac{4}{5}$ of the air which will not support burning or life is mainly nitrogen. (Lavoisier called it *azote*, which comes from a Greek word meaning 'without life'.)

In a second experiment, Lavoisier collected the red powder (mercury oxide) and heated it to a higher temperature than in the

A cartoon poked fun at Priestley's discoveries and beliefs

A painting of Lavoisier's experiment

The gases found in dry air; in the lower chart, carbon dioxide would be shown by a line 150 cm long

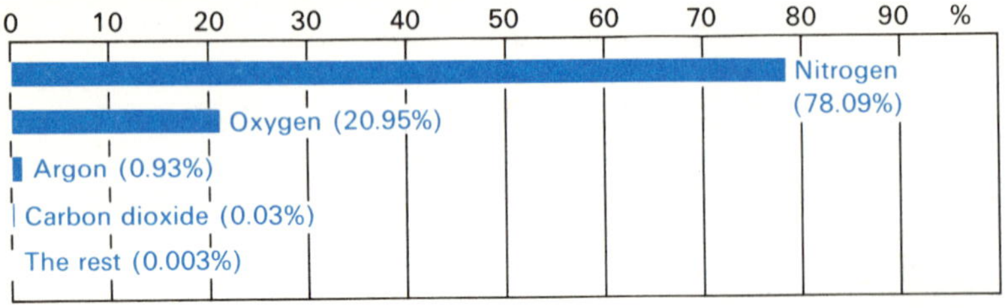

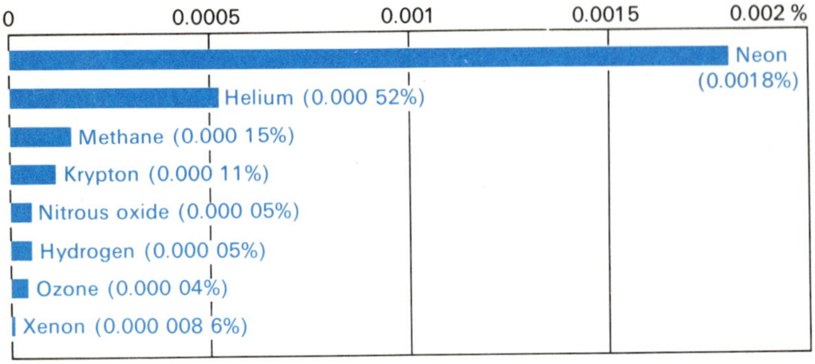

The bright lights

retort. The red powder was converted back to mercury and a gas exactly like that obtained by Priestley. The volume of gas produced was equal to the volume by which the air, in his first experiment, had been reduced. Burning substances burnt more vigorously in this gas and small animals could live in it. Lavoisier called this gas oxygen.

Lavoisier's experiments suggest:

1 Air consists of about $\frac{1}{5}$ by volume of oxygen and $\frac{4}{5}$ by volume of nitrogen.

2 When substances burn in air they combine with oxygen to form oxides.

3 Oxygen is essential for burning and life.

4 Nitrogen will not support burning or life.

Because of its unreactivity, nitrogen is used to shield artificial fibres from reaction with oxygen during their manufacture. It is also used in wine making to control the process of fermentation.

6.9 What other gases does air contain?

Accurate measurements show that dry fresh air contains 20.9% oxygen and 78.1% nitrogen. This accounts for only 99% of the air. What gases make up the remaining 1%?

THE NOBLE GASES

The presence of the noble gases (argon, helium, krypton, neon and xenon) in the air was not known until 1890. They account for nearly all the remaining 1% of dry air.

Most of the 1% is argon, with only very small amounts of helium, krypton, neon and xenon. They are also called 'the inert gases' because they are so unreactive that it is only quite recently that compounds of them have been prepared.

Helium is the most unreactive and it has a very low density. It is used to fill meteorological balloons. Unlike hydrogen, which was used previously, it will not burn.

Neon is the gas inside bright red advertising lights. When an electric spark is passed through neon a red glow is produced. The red neon glow penetrates fog, and neon lights are now used to indicate the edges of runways on airfields.

Argon and krypton are used to fill electric light bulbs. Light bulbs so filled give more light (energy) than the older vacuum ones, because they can be used at a higher temperature without the filament turning to vapour and blackening the bulb.

CARBON DIOXIDE

Carbon dioxide is formed during respiration in plants and animals, and when fuels burn. Fresh air always contains about 0.03 % carbon dioxide. Why does the percentage of carbon dioxide not increase? Because equal amounts are being removed all the time by plants in a process called photosynthesis.

Although dry fresh air contains only nitrogen, oxygen, carbon dioxide and the noble gases, ordinary air also contains variable amounts of water vapour and waste materials from industry such as sulphur dioxide, hydrogen sulphide, dust and soot.

WATER VAPOUR

The amount of water vapour in the air varies from place to place and from season to season.

At room temperature (18°C) there can be up to 4 % by volume of water vapour in the air, but only 1 % at 0°C. Why does the percentage of water vapour in the air fall at lower temperatures? Water vapour in the air comes from the evaporation of oceans, rivers and lakes; this water vapour falls back to Earth as rain. Clouds and fog consist of small droplets of water suspended in the air which form from water vapour when the temperature falls.

Dew is caused by the water vapour condensing to form water on the ground. If the temperature of this condensed water falls below 0°C, ice forms, which we call frost.

6.10 Breathing and respiration

What happens during breathing? Air is taken into the lungs and then expelled. But, what happens to it in the meantime? One way to find out is to see if air breathed in (inhaled) is any different from that breathed out (exhaled). We find that:

1 Exhaled air turns lime water cloudy more quickly than ordinary air.

2 Exhaled air produces a cloudiness on an ice-cold surface more quickly than ordinary air. The cloudiness turns anhydrous copper sulphate blue.

3 When exhaled air is breathed across the bulb of a thermometer, the temperature rises by several degrees.

Spiders' webs covered with dew . . . . .

. . . . . and frost

Try to answer these questions

1 Which gas contains more carbon dioxide, exhaled air or ordinary air?

2 What substance causes cloudiness on ice-cold surfaces?

3 Which gas contains the more of this substance, exhaled air or ordinary air?

The percentages (by volume) of carbon dioxide, water vapour and oxygen in exhaled air and in ordinary air are shown in Figure 6.6.

These experiments show that during respiration (see section 6.11):

1 oxygen is used up.
2 carbon dioxide is produced.
3 water is produced.
4 heat is produced.

Inhaled oxygen passes into the blood stream and eventually reacts with carbon and hydrogen in foods to form carbon dioxide and water.

Condensed water from exhaled breath can be used to give that extra shine

Figure 6.6

	Ordinary air	Exhaled air
Oxygen (%)	20	14
Carbon dioxide (%)	0.1	5.0
Water vapour (%)	0.0	1.1

6.11 Burning, breathing and respiration

Respiration can be summarized as:

food + oxygen ⟶ carbon dioxide + water + heat energy

This is very similar to burning (combustion) which we summarized earlier in this chapter as:

fuel + oxygen ⟶ carbon dioxide + water + heat energy

Foods are special kinds of fuel which can be used in our bodies. They contain compounds of carbon and hydrogen often combined with other elements. Fats, carbohydrates and proteins all contain large quantities of carbon and hydrogen as do obvious fuels such as petrol, coal and oil.

Respiration is a process of very slow burning. Just the same amount of heat energy is produced whether a substance burns spectacularly in the air with flames or whether it is used up slowly by our bodies as a food. The energy produced during respiration is used to enable our muscles to work and to provide heat to keep us warm.

6.12 Rusting

What happens when metals are left in contact with the air?

The shiny surface of most metals becomes dull when exposed to the air. Freshly cut sodium changes from a shiny lustrous surface to a dull white one in a matter of seconds. The white coating is a thin layer of sodium oxide. Metals such as sodium and potassium become coated with oxide very quickly and the metal below the oxide layer continus to react until all the metal is eaten away. For this reason sodium and potassium are stored under oil.

Other metals are coated much more slowly. Shiny aluminium saucepans slowly become dull as the metal reacts with air to form a thin layer of oxide which protects the metal from further reaction. Iron and steel become coated with a flaky orange-brown solid which we call 'rust'. Notice how the rust which forms flakes off very easily. This exposes a fresh surface of the metal which soon begins to rust. In time the whole metal corrodes away.

These cars were once brand new, the owners' pride

6.13 What causes rusting?

Have you noticed that your bike is more likely to rust if it is left outside in wet weather? A garden fork rusts more quickly than a pair of coal tongs kept at the fireside. It looks as though water may help rusting. What other substances take part in rusting?

A piece of sodium exposed to the air becomes coated with a layer of sodium oxide. The sodium has reacted with oxygen in the air. If rusting is similar to the reaction of sodium with air it is likely that oxygen, besides water, takes part in rusting.

Use the following experiment to decide whether oxygen and water take part in rusting. Set up five test tubes as in Figure 6.7 and leave them for several days. Place some anhydrous calcium chloride or anhydrous copper sulphate in tube 2 to absorb any water vapour and ensure that the air is dry.

The distilled water in tubes 3 and 5 must be boiled to remove any dissolved air. In tube 3, the layer of olive oil prevents air dissolving in the water. Tube 1 is the *control experiment*. It is our standard of comparison. It contains iron nails in normal moist air conditions. In many experiments it is essential to have a control experiment against which we can compare other results.

Nails in tube 1 and tube 4 rust whereas those in tubes 2, 3 and 5 do not.

Figure 6.7

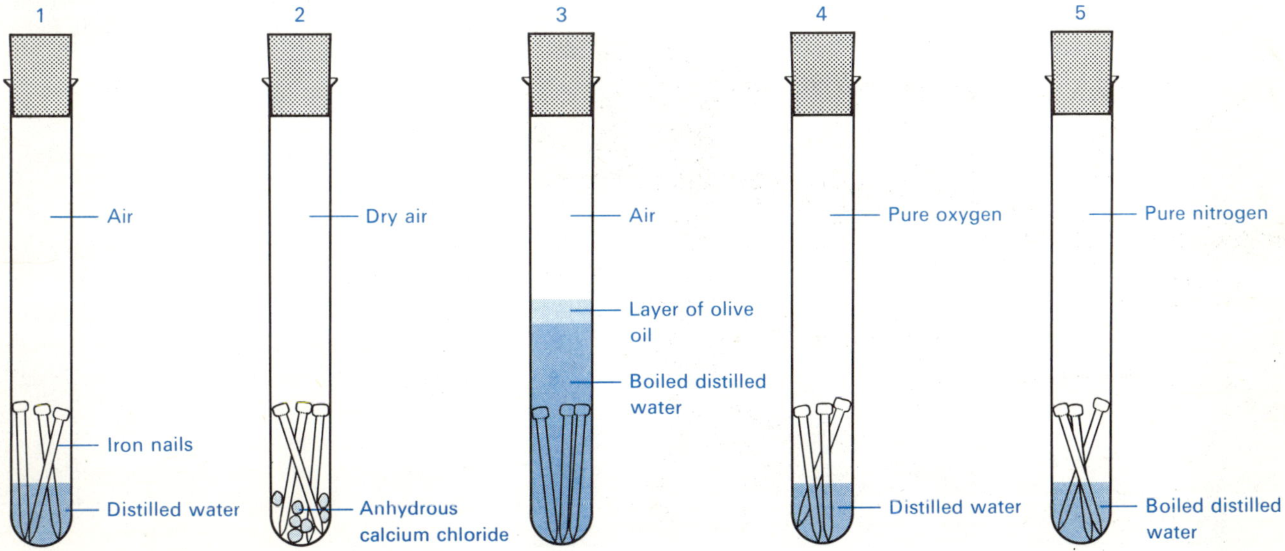

Try to answer these questions

1 Will iron rust if nitrogen is not present?
2 Will iron rust if oxygen is not present?
3 Will oxygen alone cause iron to rust?
4 Will iron rust if water is not present?
5 Will water alone cause iron to rust?

The experiment shows that both oxygen and water are necessary for iron to rust. Although rusting is a very complex process it resembles burning and respiring in requiring oxygen for it to take place.

6.14 What is rust?

Rusting, like respiration, is a process of very slow burning. No fire is visible, but during rusting iron combines with oxygen to form brown iron oxide:

$$\text{iron}_{(s)} + \text{oxygen}_{(g)} \longrightarrow \text{brown iron oxide}_{(s)}$$

The brown iron oxide combines with water to form hydrated brown iron oxide—rust. We say that substances, such as rust, that are combined with water are **hydrated**; we call them **hydrates**:

$$\underset{(brown)}{\text{iron oxide}_{(s)}} + \text{water}_{(l)} \longrightarrow \underset{(rust)}{\text{hydrated iron oxide}_{(s)}}$$

Blue copper sulphate is another example of a hydrated compound. It forms when water is added to white anhydrous copper sulphate:

$$\underset{(white)}{\text{anhydrous copper sulphate}_{(s)}} + \text{water}_{(l)} \longrightarrow \underset{(blue)}{\text{hydrated copper sulphate}_{(s)}}$$

We call the water combined in hydrates the **water of crystallization.**

A bad case of rust

6.15 The cost of rusting

Rusting costs millions of pounds every year; firstly because of the need to protect iron and steel objects, secondly because of replacement costs of rusted articles.

Although iron rusts more easily than many other metals iron and steel are used to build ships, cars, bridges and other structures because the metals are so much cheaper or stronger than other building materials. A 200 000 tonne supertanker will contain 35 000 tonnes of steel.

6.16 How is rusting prevented?

In order to prevent iron and steel rusting it is necessary to exclude water and oxygen. The most important methods of protecting iron and steel are:

1 Painting. This is the usual method of preventing rusting in objects exposed to the weather such as ships, agricultural machines, cars and bridges. Large structures, such as the Forth rail bridge, are continually being painted. As soon as the painters have finished one coat, they begin again at the other end. It takes 24 men 3 years to paint the Forth rail bridge.

2 Oiling. The moving parts of cars and other machines cannot be protected by paint, which would get scratched off. Instead they are oiled or greased.

3 Covering with a non-rusting metal. Buckets and watering cans are coated ('galvanized') with a layer of zinc by dipping them in molten zinc or by depositing the zinc by electrolysis. Other articles such as car bumpers are chromium plated by making them the cathode in an electrolytic cell.

4 Alloying. Iron and steel can be mixed with other metals to form alloys. Stainless steel contains chromium, nickel and manganese mixed with iron.

Summary

1 Air contains about $\frac{1}{5}$ by volume of oxygen and $\frac{4}{5}$ by volume of nitrogen.

2 Besides oxygen and nitrogen, air contains small amounts of carbon dioxide, noble gases, water vapour and waste materials from industry.

3 There is a similar pattern in the reactions of burning, respiration and rusting:

(a) When substances burn or rust and when animals respire they use up oxygen from the air and oxides are formed.

(b) Oxygen is essential for burning and respiring. Oxygen and water are essential for rusting.

(c) Burning, respiration and rusting cannot occur in pure nitrogen.

(d) Most fuels and foods contain compounds of carbon and hydrogen. When burning and respiration occur, the carbon and hydrogen in the food or fuel combine with oxygen in the air to form carbon dioxide and water.

4 Foods are special kinds of fuel which can be used in our bodies. Respiration is a process of slow burning.

5 When iron rusts it combines with oxygen to form brown iron oxide. The brown iron oxide reacts with water to form hydrated brown iron oxide (rust).

6 The total mass does not change during a chemical reaction (law of conservation of mass).

7 A chemical law is the summary of the results of many experiments.

It takes 24 men three years . . . . .

Study questions

1 Refer back to experiment 2 in section 6.2 (page 66).
(a) What elements does the candle certainly contain if water and carbon dioxide are produced when it burns?
(b) What other element may it contain?
(c) Why is the pump necessary?
(d) Why must the apparatus be cool before it is reweighed?
(e) Why is it necessary to repeat the experiment for the same length of time without lighting the candle if an accurate comparison is being made?

2 The Clean Air Act of 1956 forbad the emission of dark smoke from a chimney. A definition of dark smoke was made in terms of a white card with a scale composed of areas with differing degrees of black shading. New industrial plants burning solid fuel at rates of 10 tonnes per hour or more must install efficient devices to trap the grit and dust produced.
(a) How would you construct a shaded card for use in testing the darkness of smoke emitted from a chimney?
(b) How could you represent the extent of darkness of the smoke using this shaded card?
(c) What devices could industrial plants use to trap grit and dust?
(d) How would you design and locate:
(i) industrial furnaces and
(ii) industrial chimneys
to reduce smoke pollution?

3 (a) What processes cause smoke pollution in the atmosphere?
(b) Why do industrial plants get rid of their smoke through tall chimneys?
(c) What effect does smoke pollution of the atmosphere have on health? Which diseases are most linked to smoke pollution?
(d) Devise experiments:
(i) to show that city air is polluted with smoke particles.
(ii) to obtain clean air.

4 Find out what you can about air pollution. Some problems to consider are:
The effects of air pollution on health.
The effect of air pollution on animals and plants.
The cost of preventing air pollution.
The Clean Air Act (1956).
Methods of reducing air pollution.

5 What experiments would you do to investigate the effects of clean air and polluted air on the growth of pea plants?

6 In an experiment to investigate burning, a lighted candle was placed under beakers of different volume and the time was taken for the flame to go out. Each experiment was made four times. The results are given in Figure 6.8.
(a) Calculate the average times of burning in each beaker.
(b) Why was the time of burning measured four times with each beaker?
(c) Why is it important to use the same candle with all four beakers?
(d) Special beakers with flat edges (i.e., without lips) were used. Why?
(e) Plot a graph of the volume of the beaker (vertically) against the time of burning (horizontally).

Figure 6.8

Volume of beaker in cm^3	Time of burning in seconds			
450	8	7	9	8
700	15	14	16	15
1000	22	19	22	21
1200	24	28	26	26

Figure 6.9

Volume of gas removed from air	21 cm^3
Weight of silica tube+contents (at start)	11.365 g
Weight of silica tube+contents (at end)	11.393 g

Figure 6.10

Weight of test tube+red lead before heating	28.63 g
Weight of test tube+red lead after heating	28.53 g
Volume of oxygen collected	80 cm^3

Figure 6.11

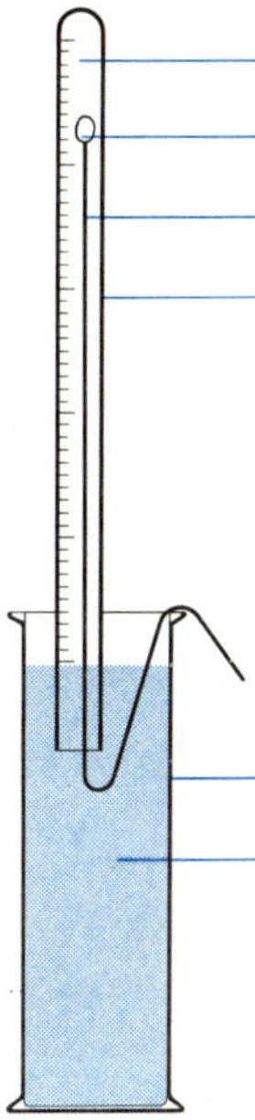

(f) How is the volume of the beaker related to the time of burning? Why is this?
(g) In what size of beaker would the candle burn for 10 seconds?
(h) In theory, the candle would burn for zero time in a beaker of its own size. What is the volume of the candle?

7 Read section 6.7 again. In the syringe experiment to determine the percentage of oxygen in the air, the hard-glass tube containing copper was weighed before and after the experiment. The results are given in Figure 6.9.
(a) How much oxygen does the copper combine with in the experiment?
(b) What causes the silica tube+contents to increase in weight?
(c) What is the weight of the oxygen removed from the air?
(d) What is the density of this oxygen? (i.e., the mass of 1 cm^3 of it).

8 In order to show that oxygen is the 'active gas' in air that combines with heated substances, a student decided to compare the density of the 'active gas' from the syringe experiment (study question 7) with the density of the oxygen evolved when red lead is heated. He weighed a sample of red lead in a test tube, heated the tube and collected the oxygen in a syringe. Finally he reweighed the test tube and its contents. His results are given in Figure 6.10.
(a) What is the weight of oxygen produced?
(b) What is the density of the oxygen? (What is the mass of 1 cm^3?)
(c) What is the density of the oxygen in grammes $litre^{-1}$?
(d) In the syringe experiment with air (study question 7) the density of the active part is 1.33 g $litre^{-1}$. How do you account for the difference in density if both gases are oxygen?
(e) Why is it important to note the temperature and pressure at which the density of a gas is measured?

9 The percentage of oxygen in the air can be determined by allowing a small piece of phosphorus to smoulder in an inverted graduated tube over water (Figure 6.11).
(a) Describe how the experiment is carried out, explaining how you would make allowance for the fact that the graduations do not go all the way down to the open end

of the graduated tube.
(b) There is no need to heat the phosphorus to make it burn. It will smoulder spontaneously in the air. What does the phosphorus react with in the air? What substance is produced? Write a word equation for the reaction taking place.
(c) Suppose you start with 50 cm^3 of air in the graduated tube. About how high will the water rise inside the graduated tube when all the oxygen has been used up? Explain how you arrive at your answer.

10 Describe an experiment which you could carry out to determine the percentage by volume of oxygen in the air breathed out by a human being. Draw a large clearly labelled diagram of the apparatus you would use, describe the measurements you would make, and explain how you would obtain a result.

11 Is air a mixture or a compound? Look back at the properties of mixtures and compounds in section 4.7. Give three reasons to support your answer.

12 In spite of his great services to mankind, Lavoisier was arrested on a trumped-up charge and guillotined on 8 May 1794. The judge is said to have pushed aside his defence with the words 'the Republic has no need of men of science'. A fellow scientist of Lavoisier is reported to have said 'Only a moment to cut off his head, and a hundred years may not give us another like it'.
(a) What great achievements did Lavoisier make?
(b) Pretend you are Lavoisier's lawyer. How would you have replied to the judge's statement that 'the Republic has no need of men of science'?

13 A sample of dry rust was heated strongly in a test tube. Droplets of liquid, *A*, formed on the cooler parts of the tube which turned anhydrous copper sulphate blue. The residue left in the tube, *B*, was heated strongly in a stream of gas, *C*. A grey-black solid, *D*, was formed (which conducted electricity) and a colourless liquid, *E*, condensed on the cooler parts of the apparatus. *E* turned anhydrous copper sulphate blue.
(a) Identify the substances, *A*–*E*.
(b) What do you think rust is?

14 In 1789, Lavoisier wrote 'In all the operations of art and nature, nothing is created: an equal quantity of matter exists before and after the change'.
(a) What general law is this called?
(b) Describe an experiment you would carry out to verify this law. Draw a diagram of your apparatus.
(c) Write a word equation for the reaction involved.

15 A student carried out a series of experiments using the metal, tin. Firstly, he weighed a sample of tin, heated it in air, and on re-weighing found that the powder he obtained weighed more than the original tin.
(a) Why does the tin increase in weight on heating?
(b) Write a word equation for the reaction which takes place.
Secondly, he heated tin in a sealed flask and noticed that, although a powder was formed, the mass of the sealed flask did not change until the seals were broken.
(c) Why does the weight remain unchanged until the seals are broken?
(d) Explain why and how the weight changes when the seals are broken.

16 Explain the following:
(a) Iron objects corrode away in time.
(b) Iron railings rust through at the bottom more quickly than at the top.
(c) The iron of shipwrecks, which has laid in deep sea-water for many years, rusts very slowly.

17 What experiments would you carry out to investigate whether:
(a) Iron rusts more quickly in sea-water, tap water or distilled water.
(b) Steel rusts more quickly than iron?
Draw a labelled diagram of the apparatus you would use in each case. Give an explanation of the results you would expect to get.

18 When aluminium is exposed to the air it becomes coated very quickly with a thin layer of oxide about 10^{-6} cm thick. This layer does not flake off, nor does it increase in thickness on standing. In order to protect the aluminium even more than its natural oxide layer does, it is possible to thicken the layer to 10^{-3} cm by a process called anodizing. The aluminium is first degreased and then anodized by making it

the anode during the electrolysis of sulphuric acid. The oxygen released at the anode combines with the aluminium and increases the thickness of the oxide layer.
(a) Why is anodizing useful?
(b) Why does aluminium not corrode away like iron, even though aluminium is coated very quickly with a layer of oxide?
(c) Write a word equation for the reaction which takes place when aluminium is exposed to the air.
(d) Carbon tetrachloride (tetrachloromethane) can be used to degrease aluminium before anodizing. Name one other liquid which would be a suitable degreasing agent.
(e) Why is water not used to degrease aluminium?
(f) Why is it necessary to degrease the aluminium?
(g) How many times thicker is the oxide coating after anodizing?
(h) Why is anodized aluminium especially useful as a building material?

19 Aluminium and iron are the two most widely used metals.
(a) Why are these two metals used so much?
(b) Iron is used in spite of the fact that it rusts easily. Mention three ways in which rusting can be prevented.
(c) What advantages are there in using iron rather than aluminium?
(d) Iron is magnetic. What uses does it have on account of this property?
(e) What advantages are there in using aluminium for certain purposes rather than iron?

Lift-off

7 Oxygen – the reactive part of air

7.1 The uses of oxygen

Oxygen is essential for burning and respiring. Burning and respiration, however, do not require pure oxygen. There are usually sufficient quantities of oxygen in the air to allow these processes to take place. But, pure oxygen is required in large quantities for industrial and medical uses. The gas Priestley first isolated in a gas jar is now produced by the tonne. For example, almost 1 tonne of pure oxygen is needed for every 10 tonnes of steel produced from impure iron. Because so much oxygen is required, modern steel works are usually equipped with an oxygen-producing plant.

Liquid oxygen is used in rockets to make the fuel burn more vigorously. The Saturn V rocket, used to lift the American astronauts into orbit during the Apollo Project, carried 2250 tonnes of liquid oxygen, 450 tonnes of kerosene and 50 tonnes of liquid hydrogen. While its jets were roaring, the first stage of Saturn V burnt 15 tonnes of kerosene–oxygen mixture per second for the first $2\frac{1}{2}$ min. This stage used up all the kerosene and 1800 tonnes of liquid oxygen. Stages 2 and 3 were powered by a mixture of liquid hydrogen and liquid oxygen.

Pure oxygen is also used in oxy-acetylene (oxygen–acetylene) welding and cutting. The mixture of oxygen and acetylene is burnt in a blow-pipe arrangement. The acetylene burns more vigorously in pure oxygen than it does in air, producing a flame with a temperature of about 3200°C. This temperature is hot enough to melt iron, steel and most other metals which can then be either cut or welded (joined together).

Pure oxygen is used in hospitals to improve respiration in patients who have lung diseases, such as pneumonia, or who have been poisoned by gases. It is also mixed with anaesthetizing gases during surgical operations.

High-altitude climbers and fliers and underwater explorers carry cylinders of oxygen–helium mixtures to assist their breathing when supplies of air are not available.

7.2 Obtaining oxygen for industry

It is important to find a large yet cheap source of oxygen in order to provide the vast amounts of pure oxygen required by industry. The cheapest and largest supply of oxygen is in the air. After all, there are about 5 000 000 000 000 000 tonnes of air in the world and 1 000 000 000 000 000 tonnes of this are

This sculptor welds together a strong lattice frame which he covers with bronze from a welding rod

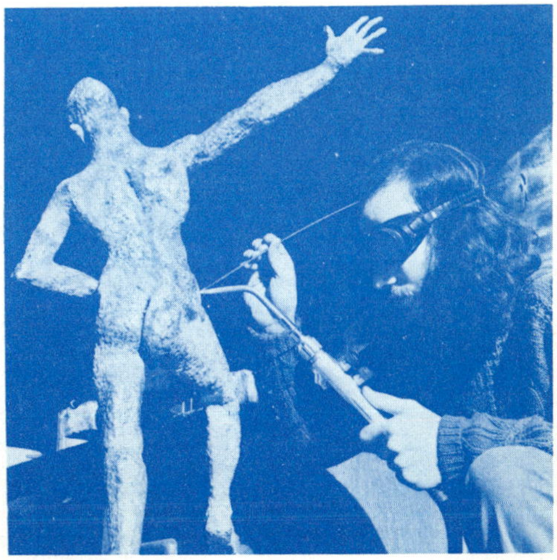

oxygen. But, how can we separate oxygen from the other gases (mainly nitrogen) in the air?

Air is first cooled to a liquid and the different constituents are then separated by fractional distillation.

Man, having no gills, has to take his own small supply of air with him for underwater swimming

Try to answer these questions

Suppose the air consists of nitrogen (b. pt. −196°C) and oxygen (b. pt. −183°C).

1 When the temperature of air falls slowly, which gas (oxygen or nitrogen) liquefies first?

2 When liquid air (at −200°C) is allowed to warm up slowly which gas (nitrogen or oxygen) forms first?

In industry, air is cooled to about −200°C to obtain liquid air. Figure 7.1 shows a diagram of the processes involved in the manufacture of liquid air.

The air is first cleaned of dust and other solid impurities and then compressed. When a gas is compressed it gets hot, as a bicycle pump gets hot when you blow up tyres. The air is passed through a water-cooled coil to cool it back to its original temperature.

The air is then allowed to escape rapidly from a jet in the expansion chamber. This sudden expansion causes cooling. Have you noticed the cooling produced when compressed carbon dioxide escapes from a fire extinguisher? The air escaping from the jet is colder than that inside the coiling in the expansion chamber. Thus, the air approaching the jet inside the coiling is cooled by the air which has already escaped, and it is cooled still more when it escapes from the jet itself.

The cooled air circulates around the apparatus continuously and in time the temperature is low enough to liquefy the air, which collects below the jet. The liquid air is mainly liquid nitrogen (b. pt. −196°C) and

Figure 7.1

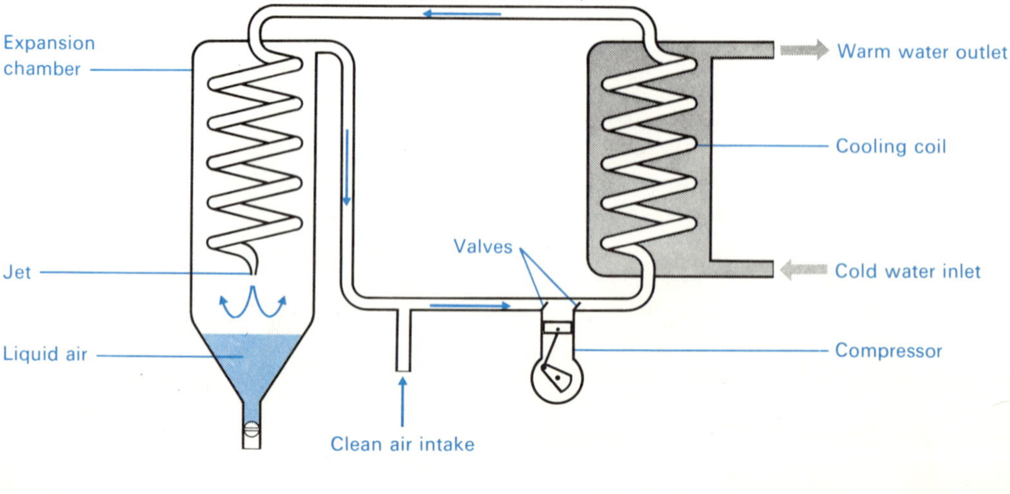

liquid oxygen (b. pt. $-183°C$). The liquid air is now removed from the chamber and separated into pure oxygen and pure nitrogen by fractional distillation.

When liquid air is warmed slowly, liquid nitrogen turns to a gas at a lower temperature than liquid oxygen. The nitrogen boils off first, leaving behind the liquid oxygen. The liquid oxygen can be stored as a liquid or converted to a gas and stored under pressure in steel cylinders.

The other gases in the atmosphere (particularly the noble gases) are also obtained during this process. They too can be stored in cylinders.

7.3 Obtaining oxygen in small quantities in the laboratory

It is not usually necessary to prepare oxygen because cylinders of oxygen are so cheap. However, oxygen can be prepared if necessary by decomposing compounds which contain oxygen, such as red lead, potassium permanganate and red mercury oxide. Priestley originally produced oxygen by heating red mercury oxide:

$$\underset{(red)}{\text{mercury oxide}_{(s)}} \xrightarrow{heat} \text{mercury}_{(l)} + \text{oxygen}_{(g)}$$

Oxygen is not prepared from red mercury oxide nowadays. A lot of red mercury oxide only gives a little oxygen. Anyway, red mercury oxide is very expensive and mercury vapour is poisonous.

A much cheaper and more convenient method of preparing oxygen is by decomposing hydrogen peroxide using manganese dioxide. A suitable arrangement of apparatus for preparing and collecting the oxygen is shown in Figure 7.2.

Add the aqueous solution of hydrogen peroxide slowly to the solid manganese dioxide in the round-bottomed flask. As soon as the hydrogen peroxide touches the manganese dioxide there is a vigorous reaction. Hydrogen peroxide decomposes and oxygen is produced:

$$\text{hydrogen peroxide}_{(aq)} \longrightarrow \underset{(water)}{\text{hydrogen oxide}_{(l)}} + \text{oxygen}_{(g)}$$

The first gas jar to be filled with gas will contain mainly air, displaced from the flask and the delivery tube. Oxygen is collected over water even though slightly soluble in water. If it wasn't soluble, fishes and other aquatic animals which use this dissolved oxygen during respiration would die.

7.4 Catalysts

Hydrogen peroxide solution decomposes very slowly at room temperature and a little faster when it is warmed. It decomposes much faster when manganese dioxide is present. Manganese

Figure 7.2

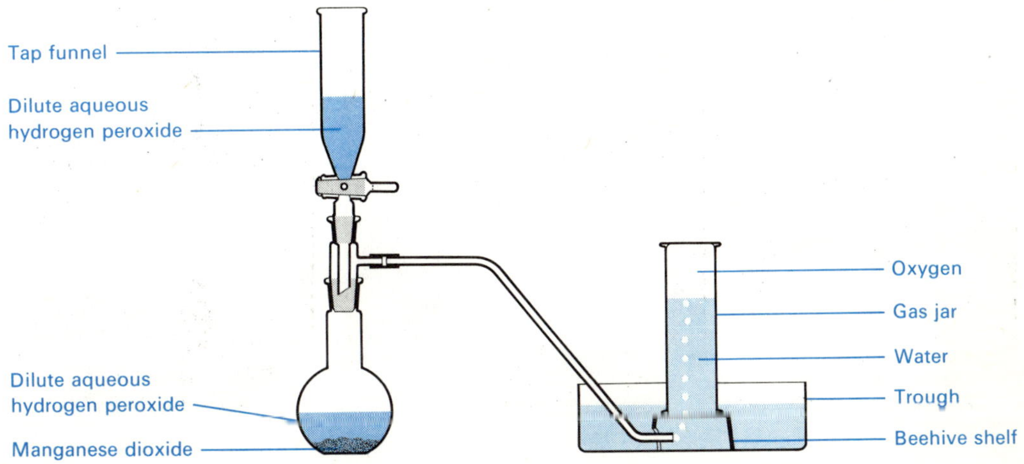

dioxide helps the hydrogen peroxide to decompose more easily. However, manganese dioxide is not used up during the reaction. The manganese dioxide left at the end weighs exactly the same as that present at the beginning. The manganese dioxide has not changed chemically during the decomposition of the hydrogen peroxide, yet it has made decomposition much easier. The manganese dioxide has acted as a **catalyst** for the reaction.

The two important properties of catalysts are:

1 They change the rates of chemical reactions.

2 They are not changed chemically during reactions.

A reaction in which a catalyst takes part is called a **catalysis.**

Catalysts play an important part in the chemical industry. Sulphuric acid, nitric acid and ammonia are produced by processes involving catalysts. Each of these chemicals is used in large quantities to produce other materials for industry and agriculture.

Catalysts make reactions more efficient (the products are obtained more quickly), thus reducing the costs of manufacture.

One of the most important discoveries of this century was that all chemical processes in living things require catalysts. These catalysts which make possible biological processes are called **enzymes.** Enzymes in our bodies are responsible for breaking down the foods we eat to provide us with energy for warmth and muscular and mental work. Enzymes are also responsible for building up the chemicals which make up our bodies; proteins in our muscles, carbohydrates and fats which are stored to provide energy when required and DNA (deoxyribose nucleic acids) in our genes.

Liver contains the enzyme catalase: catalase (and hence raw liver) will decompose hydrogen peroxide faster than manganese dioxide will.

Most reactions which are catalysed by enzymes are impossible to copy in the laboratory and are still beyond the ability of the world's leading chemists.

Think about these questions

1 See if other substances (e.g., copper or sand) besides manganese dioxide catalyse the decomposition of hydrogen peroxide.

2 How does manganese dioxide speed up the decomposition of hydrogen peroxide?

3 How do catalysts change the rates of chemical reactions?

Try to answer these questions

Which of the following statements about catalysts are true and which are false?

Catalysts:

1 may be solids, liquids or gases.

2 may slow reactions down.

3 increase the amount of product formed from a given amount of reactant.

4 are used up in the reactions they catalyse.

5 are chemically the same at the end of the reaction as at the start.

Liver decomposes hydrogen peroxide faster than manganese dioxide will (*far right*)

7.5 How does oxygen react with elements?

We already know something about the reactions of elements with oxygen in the air. Iron is slowly covered with a layer of brown rust (iron oxide) and a shiny aluminium surface is covered rapidly with a dull white layer of aluminium oxide. We have also noticed that some elements react with oxygen more vigorously than others.

Put a piece of an element into a deflagrating spoon (Figure 7.3) and heat it strongly until it starts to burn. Then place the deflagrating spoon into a gas jar containing oxygen. Figure 7.4 lists the information obtained when various metal and non-metal elements were burnt in pure oxygen.

Figure 7.3

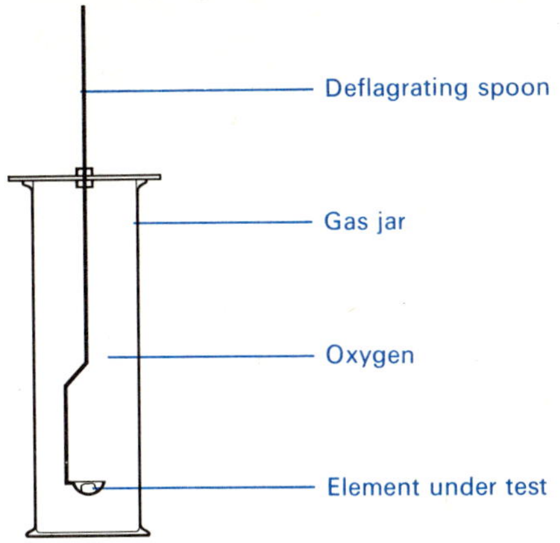

Figure 7.4

Element	Description of burning	Product of burning	Colour of universal indicator with solution of product	pH of universal indicator with solution of product
Carbon	Red hot carbon (charcoal) glows brightly in oxygen	Colourless gas – carbon dioxide	Orange-yellow	5
Sulphur	Molten sulphur burns strongly in oxygen with a bluish flame	Colourless gas – sulphur dioxide	Orange	3
Phosphorus	Molten phosphorus burns very brightly in oxygen with a yellow flame producing thick white smoke	White smoke and powder – phosphorus pentoxide	Orange	2
Sodium	Sodium burns with a bright yellow flame. Clouds of white smoke are produced which settle as a white powder	White solid – sodium oxide	Blue	11
Magnesium	Magnesium burns with a dazzling white flame. White clouds of smoke are produced which settle to a white powder	White solid – magnesium oxide	Greenish-yellow	8
Iron	Fine iron wire burns with a shower of sparks producing a black-brown solid	Black-brown solid – iron oxide	The product is insoluble and has no effect on universal indicator	
Copper	No flame or sparks. The copper becomes covered with a black solid	Black solid – copper oxide	The product is insoluble and has no effect on universal indicator	

Hot iron filings react with chlorine to form clouds of brown iron chloride

WHAT SUBSTANCES ARE PRODUCED?

All the elements burn more vigorously in oxygen than in air. The substances produced when elements burn in oxygen are oxides:

$$\text{carbon}_{(s)} + \text{oxygen}_{(g)} \longrightarrow \text{carbon dioxide}_{(g)}$$

$$\text{sulphur}_{(s)} + \text{oxygen}_{(g)} \longrightarrow \text{sulphur dioxide}_{(g)}$$

$$\text{sodium}_{(s)} + \text{oxygen}_{(g)} \longrightarrow \text{sodium oxide}_{(s)}$$

$$\text{magnesium}_{(s)} + \text{oxygen}_{(g)} \longrightarrow \text{magnesium oxide}_{(s)}$$

All the metal oxides (those of sodium, magnesium, iron and copper) are solids. Two of the non-metal oxides (carbon dioxide, sulphur dioxide) are gases and one (phosphorus pentoxide) is a solid.

In the same way, substances produced by combination of elements with sulphur are called **sulphides** and substances produced by combination of elements with chlorine are called **chlorides.**

Sulphides can be made by heating solid elements with sulphur, and chlorides can be made by heating solid elements in a stream of chlorine.

For instance, iron filings mixed with powdered sulphur and heated react exothermically to form a grey-black solid, iron sulphide:

$$\text{iron}_{(s)} + \text{sulphur}_{(s)} \xrightarrow{\textit{heat}} \text{iron sulphide}_{(s)}$$

Hot iron filings placed in a gas jar of chlorine form brown clouds of iron chloride:

$$\text{iron}_{(s)} + \text{chlorine}_{(g)} \xrightarrow{\textit{heat}} \text{iron chloride}_{(s)}$$

All the oxides, sulphides and chlorides of metals are solids at room temperature, but the oxides, sulphides and chlorides of non-metals are usually gases or liquids (phosphorus pentoxide is an exception to this general rule).

In general, metals react with non-metals to form solids, non-metals react with non-metals to form liquids or gases, but metals do not react with each other. However, mixtures of metals, called alloys, have important commercial uses.

WHAT IS THE ORDER OF REACTIVITY?

Some metals, such as sodium and magnesium, react very vigorously (and exothermically) with oxygen producing a bright flame. Others, such as copper, react very slowly without producing any flame. These metals only react to form a thin layer of oxide. We can arrange the metals in decreasing order of reactivity so that the most reactive metal is at the top of the series and the least reactive is at the bottom.

This series is called the **activity series** of metals. Figure 7.5 shows an activity series for the four metals we used. This is only a very rough order at present. It is difficult to make a clear distinction between some elements based on their reaction with oxygen (e.g., sodium and magnesium).

The form in which the metals are used will also make a difference. Metals (and other things) will burn more easily when they are in the form of fine wire or powder or in thin sheets than when they are in large pieces. For example, iron filings will react more easily and more vigorously than a large chunk of iron. Try lighting a fire with a tree trunk!

Nevertheless, there are some obvious differences in reactivity. Sodium, calcium and magnesium burn fiercely in oxygen with very bright flames producing clouds of white smoke. Zinc and iron burn less vigorously; copper and lead do not burn with any flame or sparks but only form a layer of oxide on their surface. Some metals such as gold and platinum are so unreactive that they are unchanged, and quite shiny, even after being heated in pure oxygen.

DO THE PRODUCTS REACT WITH WATER?

In each case, the substance produced on burning was shaken with distilled water and the solution produced was then tested with universal indicator paper. The results are given in Figure 7.4. Certain substances, such as universal indicator, can tell us whether a solution is acidic.

These substances, called **indicators,** change colour depending on how acidic the solution is in which they are placed.

It would be very clumsy to refer to the colour of the indicator (to describe how acidic a substance is), so chemists use a scale of numbers, known as the **pH scale.** (Figure 7.6). On this scale, a pH below 7 indicates that a substance is **acidic,** and a pH above 7 indicates that a substance is **alkaline.** Solutions with a pH value of 7, such as pure water, are neither acidic nor alkaline. They are **neutral.** Universal indicator shows a slightly different colour for

Figure 7.5

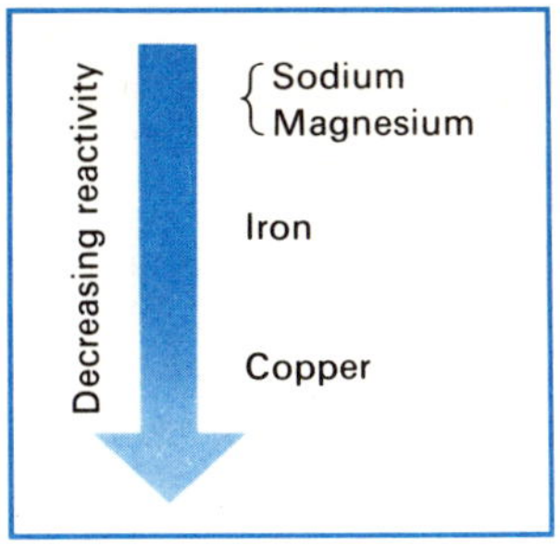

Figure 7.6

pH		Colour of universal indicator
14	Very alkaline	Violet
13		
12		Blue
11	Increasing alkalinity	
10		Bluish green
9	Slightly alkaline	Green
8		Greenish yellow
7	Neutral	Yellowish green
6	Slightly acid	Yellow
5		Orange-yellow
4		
3	Increasing acidity	
2		Orange
1	Very acid	
0		Red

Figure 7.7

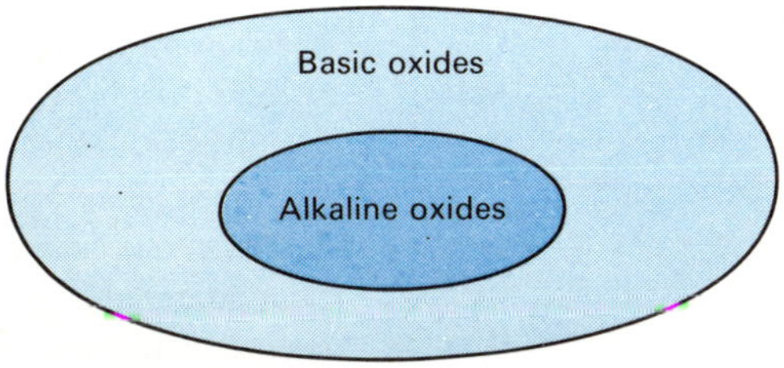

each value of pH. Thus, the colour of the indicator tells us the pH of the solution (i.e., how acid or alkaline it is).

Figure 7.4 listed the colours of the solutions produced and their corresponding pH values when the oxides are dissolved in water and universal indicator is added.

The three non-metals (carbon, sulphur and phosphorus) produce oxides which react with water to form solutions with pH values less than 7. Substances which dissolve in water to form solutions with pH values of less than 7 are called **acids.** These solutions give a red or orange colour with universal indicator:

carbon dioxide$_{(g)}$ + water$_{(l)}$ $\longrightarrow$ carbonic acid$_{(aq)}$

sulphur dioxide$_{(g)}$ + water$_{(l)}$ $\longrightarrow$ sulphurous acid$_{(aq)}$

phosphorus pentoxide$_{(s)}$ + water$_{(l)}$ $\longrightarrow$ phosphoric acid$_{(aq)}$

The oxides of non-metals which give acids in water are called **acidic oxides.**

Sodium and magnesium oxides react with water to form alkaline solutions whose pH values are greater than 7. Such solutions give a blue or green colour with universal indicator:

sodium oxide$_{(s)}$ + water$_{(l)}$ $\longrightarrow$ sodium hydroxide$_{(aq)}$

magnesium oxide$_{(s)}$ + water$_{(l)}$ $\longrightarrow$ magnesium hydroxide$_{(aq)}$

Substances, such as sodium hydroxide and magnesium hydroxide, which dissolve in water to form alkaline solutions with a pH greater than 7 are called **alkalis.**

Iron oxide and copper oxide do not react with water, nor do they dissolve in water. Thus, they do not affect the pH of water and do not change the colour of universal indicator.

The oxides of metals are called **basic oxides.** Most metal (basic) oxides are insoluble in water, but a few react with water to form alkalis. Those that do are sometimes called **alkaline oxides.** Figure 7.7 represents the relationship of basic oxides and alkaline oxides in a Venn diagram.

Try to answer this question

Barium burns in oxygen to form a white oxide. This oxide dissolves in water to form an alkaline solution. Which of the following

Figure 7.8

Substance	Acid present
Acid drops	Citric acid
Lemon juice	Citric acid
Citrus fruit (oranges, lemons, grapefruit)	Citric acid
Vinegar	Acetic acid
Sour wine	Acetic acid
Sour milk	Lactic acid
Soda water	Carbonic acid
Rhubarb	Oxalic acid
Ant stings	Formic acid

pH values of household substances

pH	Substance
14	Oven cleaner
13	
12	
11	Washing soda Bleach (Domestos)
10	Quicklime
9	
8	Indigestion powder Toothpaste
7	Salt Milk
6	Acid drops Sour milk
5	Lemon juice
4	Vinegar
3	
2	
1	Hydrochloric acid

elements resemble barium in this respect?

Copper	Sulphur	Phosphorus
Sodium	Calcium	Magnesium
Iron	Carbon	Potassium

PATTERNS IN REACTION

These results emphasize the different patterns in reaction of metals and non-metals. In general:

1 Metals react with oxygen to form solid oxides, whereas non-metals usually react with oxygen to form gaseous oxides.

2 Metal oxides are either insoluble or react with water to form alkalis, whereas non-metal oxides react with water to form acids.

A good method of telling whether an element is a metal or a non-metal is to produce the oxide and see whether the oxide reacts with water to form an acid, an alkali or is insoluble.

Although there are a few exceptions (such as phosphorus pentoxide, a solid non-metal oxide) this general pattern in reaction shows yet again that a simple division of elements into metals and non-metals is a useful one.

7.6 Using indicators to detect acids and alkalis

Non-metal oxides react with water to produce acids, but what other substances are acids? What acidic and alkaline substances do we use in everyday life? Figure 7.8 gives a list of acid substances and the acids they contain. Acidic substances usually have a sour or sharp taste; acid drops, lemon juice and vinegar all contain acids. Milk and wine go sour because of the production of acids in them.

However, it would be very hazardous to rely on taste as a means of detecting acids! Prussic acid (hydrogen cyanide) is extremely poisonous, and concentrated sulphuric and nitric acids will burn the skin. For this reason it is useful to have indicators to determine the acidity of substances.

The first indicators which chemists used were naturally occurring coloured substances, such as dyes in vegetables, fruit and flowers. The dyes from blackcurrants, rose petals and red cabbage are very effective in testing for acidity and alkalinity. Almost any coloured plant material is suitable, except yellow flowers such as dandelions and daffodils. Litmus, an indicator which is still used nowadays, was originally extracted from lichen.

Unfortunately, these natural coloured substances cannot tell us how acid or how alkaline something is. They are the same colour in a very weak acid solution as in a very strong acid solution. Also, the colour does not change from weakly alkaline to strongly alkaline solution. For example, litmus is red in all acid solutions with pH less than 7 and blue in all alkaline solutions with pH greater than 7.

Chemists have made indicators which can tell us accurately how acid a substance is. These indicators which show the degree of acidity or alkalinity are more useful and more effective than the natural coloured substances. Universal indicator is a mixture of several compounds which have been purified or synthesized by chemists (Figure 7.6).

The left-hand plot has had no treatment since 1863; the pH value of the soil is above 5. The other plot has been treated with ammonium sulphate, a fertilizer that gradually makes the ground acidic; the pH value of this plot is below 4

Figure 7.9

Number of measures of calcium hydroxide added	Temperature of resulting solution in °C	pH of resulting solution
0		
1		
2		

7.7 Neutralization

IN THE GARDEN

Most plants, especially vegetables, grow best in neutral or slightly acid soil between pH 6·5 and 7·0. Below pH 6·5 the soil is too acid ('sour') and is harmful to most plants. Some heathers and rhododendrons grow best in more acid soils of pH less than 6·5. The soil can be 'cured' of its acidity, as farmers say, by adding quicklime which removes the acidity.

Quicklime (calcium oxide) reacts with water in the soil to form an alkali (calcium hydroxide). Farmers check the pH of soil by shaking a sample of it with water and indicator, and comparing the colour of the solution against a chart.

In the following experiment, you can follow the pH change of an acid solution as small amounts of alkali are added to it. Use 10 cm^3 of clear vinegar diluted with 10 cm^3 of water as your acid solution.

Withdraw 1 drop of solution on the end of a stirring rod and place it on a piece of universal indicator paper. Compare the colour with the indicator chart and record the corresponding pH. At the same time record the temperature of the acid solution.

Add 1 measure of calcium hydroxide (about $\frac{1}{5}$ of a flat teaspoonful, or about 0·05 g) to the acid solution, stir, and test the pH and the temperature of the resulting solution as before. Continue to add further measures of calcium hydroxide, testing the pH and temperature after each addition, until the colour of the indicator no longer changes.

Record your measurements in a table as shown in Figure 7.9.

Figure 7.10

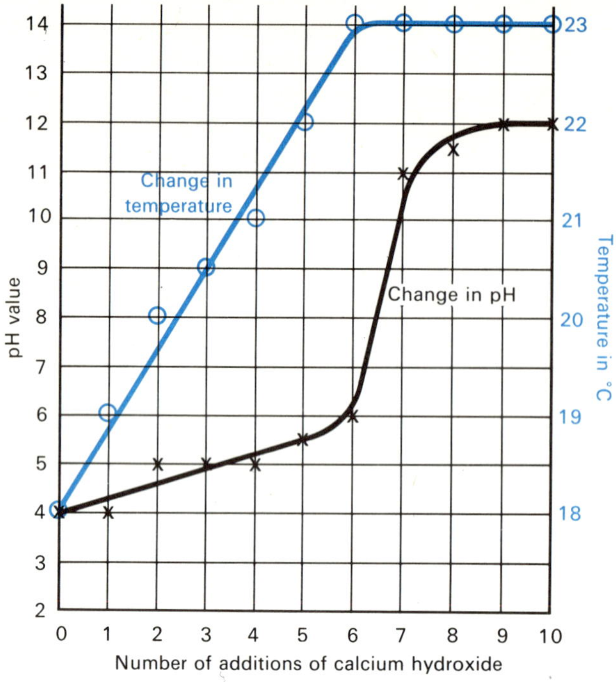

Figure 7.10 shows the change in pH and temperature with each addition of calcium hydroxide.

Try to answer these questions

1 Why does the temperature rise for the first six additions of calcium hydroxide?

2 Why does the temperature rise about the same amount for each addition of calcium hydroxide?

3 Why does the temperature stay constant after the sixth addition?

4 Why does the pH change rapidly from 6 to 11 when the seventh addition of calcium hydroxide is made?

5 Why does the pH change very little after the seventh addition?

Each of the first six additions of calcium hydroxide causes a temperature rise of about 1°C. The increase in temperature shows that a chemical reaction is taking place between the acid and the alkali. This reaction is called **neutralization** because the acidic solution is made neutral by the reaction of the acid with alkali.

If excess calcium hydroxide is added, the solution passes through neutral and becomes alkaline, and there is a large change in pH value. Between six and seven measures of alkali are needed to neutralize the acid.

By doing this sort of experiment on soil, farmers can measure how much quicklime they need to neutralize the acidity of the soil.

Basic compounds, such as magnesium hydroxide, neutralize the excess of acid in the stomach

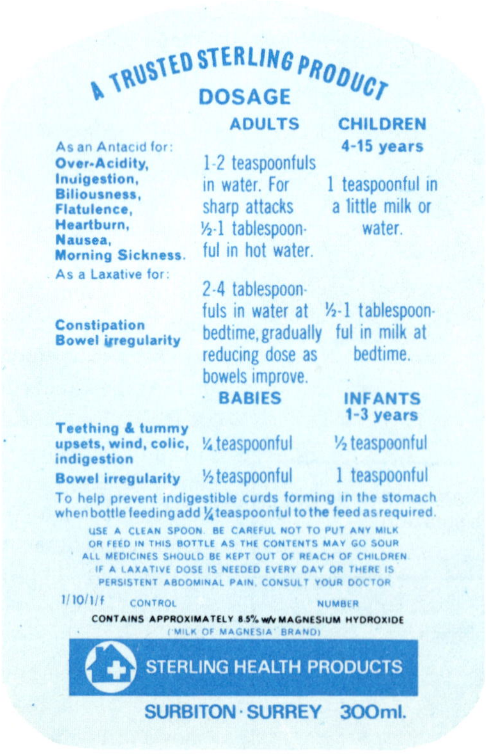

IN THE HOME

Advertisements for indigestion cures talk of 'acid stomach' and 'acid indigestion'. These medicines usually consist of basic compounds which neutralize the excess of acid during stomach ache and relieve indigestion and pain.

The process of neutralizing acidity is also important in cleaning your teeth. Teeth are washed by saliva which is very slightly alkaline and acts as a lubricant. The saliva contains millions of bacteria. Every time you eat sugar, bacteria in your mouth feed on it and convert it very quickly to acids which 'eat into' the enamel of teeth and cause decay. The saliva eventually neutralizes these acids produced from the sugar. Teeth are not attacked by the alkalis in saliva. Toothpaste is also alkaline (pH about 8) and cleaning your teeth with toothpaste helps to neutralize the acidity more quickly.

Figure 7.11 shows how the pH in your mouth changes throughout the day.

The light blue area represents the saliva when it is alkaline and the teeth are not under attack from acids.

The darker blue area represents the saliva when it is acidic and the teeth are under attack from acids. Notice that the saliva is acidic during and just after meals when sugar in the food is being converted into acids by bacteria in the saliva.

The blue line represents the pH in your mouth from waking to bed-time. If you eat four regular meals each day, with no sweets or snacks in between, there are only four possible times when acids in your mouth can attack the enamel of your teeth and cause decay. Most of the time the pH in your mouth is alkaline and your teeth are safe from attack.

Now look at Figure 7.12. In addition to four meals, you have sucked a sweet on the way to school, had a mid-morning snack, biscuits or lollipops during the afternoon and further snacks while watching television in the evening. Your saliva has been acidic during more of the day and your teeth will have been under frequent attack by acids. You can see from the graph how important it is:

1 to eat sweets only after a meal and not between meals.

2 to brush your teeth regularly with toothpaste, especially after breakfast.

Figure 7.11

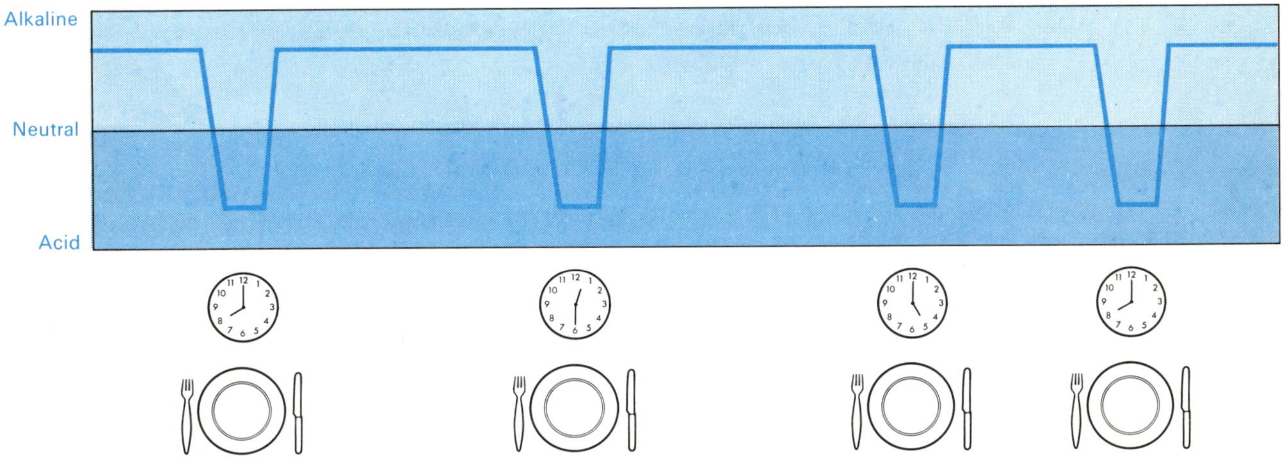

Figure 7.12

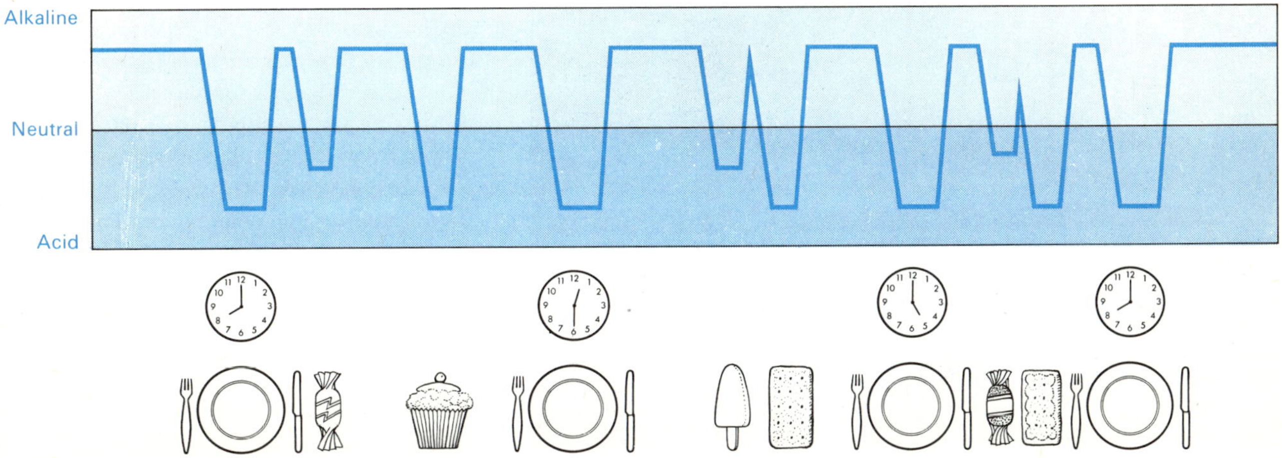

IN INDUSTRY

The chemical industry uses large quantities of sulphuric acid, nitric acid and hydrochloric acid. Sulphuric acid is one of the world's most important chemicals. More then 30 million tonnes of it are used each year. About one quarter of the sulphuric acid is used in the manufacture of 'superphosphate' fertilizer from natural phosphates.

It is also used in the manufacture of ammonium sulphate for fertilizers, copper sulphate for fungicides and in the manufacture of rayon, drugs and dyes. The uses of sulphuric acid are so numerous that the importance of an industrial nation can be judged by the quantity of sulphuric acid it produces. The wide use of sulphuric acid is due to the fact that in most cases it is the cheapest and most easily available strong acid.

Nitric acid and hydrochloric acid also have many uses but they are less important than sulphuric acid. Nitric acid is used chiefly to make explosives, such as TNT and dynamite, and fertilizers, such as ammonium nitrate. Large quantities of ammonium nitrate are mixed with chalk and sold under trade names such as 'Nitrochalk'.

Hydrochloric acid is used industrially for cleaning metal surfaces, particularly steel, before it is coated with zinc or tin to make galvanized steel or tin plate.

The two most important industrial alkalis are sodium hydroxide (commonly called caustic soda) and calcium hydroxide (commonly called slaked lime).

Caustic soda is used in the manufacture of soap from fats, in 'mercerizing' cotton fibres to give them a glossy appearance, in refining petroleum, and in the production of cellulose and rayon.

Slaked lime is prepared by adding water to quicklime (calcium oxide). It is widely used as an industrial alkali. It is also used in the production of caustic soda, in bleaching powder and it is a constituent of mortar. Quicklime is used in the manufacture of calcium carbide, from which acetylene is obtained.

Soap-making: oils and a measured amount of alkali are steam-heated. A vigorous reaction takes place which changes the oils into soap. The man with the trowel is gauging from the colour, thickness, stickiness, etc., of the mixture, and from his own experience, whether the reaction is complete

Summary

1 Oxygen is obtained industrially by the fractional distillation of liquid air.
2 If required in small quantities, oxygen can be prepared by the decomposition of oxygen-containing compounds.
3 Catalysts change the rate of a chemical reaction, but are themselves unchanged chemically as a result of the reactions they catalyse.
4 The compounds of metals (oxides, sulphides and chlorides) are usually solids at room temperature, whereas those of non-metals are usually gases or liquids.
5 The metals can be arranged in an activity series to show their relative vigour of reaction with other substances.
6 Metal oxides are either insoluble or react with water to form alkalis, whereas non-metal oxides react with water to form acids.
7 Indicators change colour depending on the acidity or alkalinity of the solution they are in.
8 Acids and acidic solutions have a pH less than 7. Alkalis and alkaline solutions have a pH greater than 7.
9 Neutralization occurs when an acid is made neutral by reaction with an alkali.

Study questions

1 A student attempted to prepare and collect oxygen using the apparatus shown in Figure 7.13.
(a) Name the solid and the liquid which would be suitable for making oxygen in this apparatus.
(b) Name the piece of apparatus labelled *A*.
(c) Write a word equation for the reaction which takes place.
(d) Having made a suitable choice of reagents, the student found that although a gas was obviously being formed, none of it was collecting in the test tube, *B*. Explain why this was so. How would you overcome the fault, using the same apparatus and materials?
(e) How would you show that the gas collected in tube *B* contained oxygen?
(f) What experiments could you carry out to find the percentage of oxygen in the sample?

2 A student was asked to investigate the effect of quicklime on the production of tomatoes. He mixed his soil evenly and then divided it equally. With one half of the soil he mixed a weighed amount of quicklime. Using the two samples of soil, he planted 12 tomato plants in the 'unlimed' soil and 12 tomato plants in the 'limed' soil. He then treated both groups of plants identically.
(a) Why do you think he went to the trouble of planting 12 tomato plants in 'unlimed' soil?
(b) Why did he go to the trouble of having 12 'limed' plants and 12 'unlimed' ones, instead of just one of each?

3 Four substances, *A*, *B*, *C* and *D*, were heated in oxygen. They behaved as follows:
A melted, but did not form a new substance.
B burned to form carbon dioxide and water.
C burned to form a single oxide.
D did not melt and did not form a new substance.
(a) Which substances may be metals?
(b) Which substances may be elements?
(c) Which substances may be compounds?
(d) Which of the four substances will the following resemble?

Carbon	Petrol	Salt
Sulphur	Paper	Gold
Bread	Glass	Polythene

4 A student accidentally spilt some concentrated sodium hydroxide on his clothes. He should add something (which must not do further damage) to neutralize the alkali.
(a) Which of the following four substances would be most useful?
Vinegar (pH = 4).
Orange juice (pH = 6).

Figure 7.13

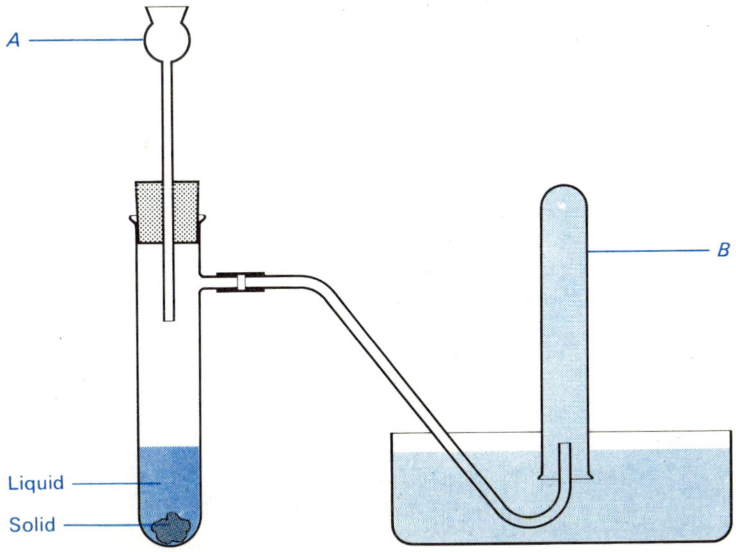

Sodium bicarbonate solution (pH = 8).
Potassium hydroxide solution (pH = 13).
(b) Explain clearly why you chose the particular substance.
(c) Which of the four substances in part (a) would you use as an eyewash when the following has been splashed in someone's eye?
(i) acid.
(ii) alkali.
Explain your choice.

5 Below are seven questions which relate to the substances *A*–*D* in Figure 7.14.
(a) Which substance melts if put into a test tube surrounded with boiling water?
(b) Which of the substances *A*, *B* or *C* is a liquid over the greatest temperature range?
(c) Which substance could be a metal?
(d) Which substance could be a non-metal?
(e) Which substance is most like sodium chloride?
(f) Which substance is most like sulphur?
(g) What elements does *D* contain?

6 Quicklime (calcium oxide) or slaked lime (calcium hydroxide) are added to the soil in many areas to neutralize the acidity and improve the vegetation. Why is the addition of calcium valuable to livestock (particularly cattle which feed on the vegetation) and to mankind?

7 (a) How is oxygen produced industrially?
(b) How would you determine experimentally the percentage of oxygen present in air?
(c) Why do explosive compounds and mixtures, such as gunpowder, contain a large percentage of oxygen?
(d) Why does the percentage of oxygen in the air not change?

8 (a) You are told that when a certain black insoluble powder catalyses the decomposition of hydrogen peroxide, the same amount of black powder is present after the reaction is complete as there was at the beginning of the reaction. How would you verify this?
(b) Figure 7.15 concerns an experiment to investigate the production of oxygen from hydrogen peroxide:
(i) in an uncatalysed reaction.
(ii) in a catalysed reaction.
The amount of hydrogen peroxide taken was the same in each experiment. The volume of oxygen produced was recorded at intervals up to 24 min after the start of the reaction in each case.
Record the results of these two experiments on the same graph by plotting the volume of oxygen produced (vertically) against time (horizontally).
(c) Write a word equation for the reaction involved.
(d) During which one of the following 2 min intervals was the catalysed reaction quickest?

0–2 min	12–14 min
4–6 min	20–22 min
8–10 min	

(e) Indicate on your graph what would happen in the next 4 min if:
(i) more catalyst was added to the flask containing the catalysed reaction after

Figure 7.14

Substance	M. pt. in °C	B. pt. in °C	Electrical conductivity	Effect on heating
A	1540	3000	Good when solid or liquid	Burns to form a single oxide which is insoluble in water
B	44	280	Very poor	Burns to form a single oxide which forms an acidic solution with water
C	770	1407	Very poor when solid but good when liquid	Melts, but no new substance forms
D	Begins to melt at about 200	Reacts before b. pt.	Very poor when solid or liquid	Melts then burns to form carbon dioxide and water

24 min.
(ii) some of the catalyst was added to the flask containing the uncatalysed reaction after 24 min.

9 What colour does dry blue litmus paper turn when placed in separate samples of:
(a) dilute hydrochloric acid.
(b) salt solution.
(c) moist carbon dioxide.
(d) sugar solution.
(e) orange squash.
(f) dry carbon dioxide?

10 Using the apparatus in question 1 explain how you would investigate the minimum quantities of solid manganese dioxide and aqueous hydrogen peroxide required to produce one test tube full of oxygen at atmospheric temperature and pressure. Give some indication of the results you would expect to get with your experiments.

11 Describe experiments and the reasoning behind them which show that water:
(a) is formed when hydrogen combines with oxygen.
(b) can be decomposed to form hydrogen and oxygen.
(c) is not a mixture of hydrogen and oxygen.

Figure 7.15

Time in minutes	Volume of oxygen in cm^3	
	(i)	(ii)
0	0	0
2	1	10
4	2	23
6	6	40
8	10	85
10	16	175
12	22	220
16	35	246
20	50	250
24	66	250

12 Indicate which of the following are solids, which are liquids and which are gases at room temperature and pressure:

Nitrogen	Argon
Molybdenum oxide	Zinc chloride
Hydrogen oxide	Tungsten
Nitrogen oxide	Magnesium sulphide

13 Which of the following is (are) true?
A catalyst is a substance that is:
(a) necessary for a particular reaction to take place.
(b) used to ensure the products of a reaction are pure.
(c) always a solid.
(d) sometimes a metal.
(e) used to change the speed of a chemical reaction.
(f) used to make a reaction less dangerous.
(g) sometimes used to slow reactions down.

14 (a) Write word equations to summarize the reactions which take place when the following react with oxygen:

Calcium	Carbon
Zinc	Sulphur

Into what two groups can the products be divided?
(b) Which of the following are gases and which are solids at room temperature and pressure?

Nitrogen dioxide	Chromium oxide
Tin oxide	Chlorine monoxide

15 Two companies manufacture tablets of identical size which are reputed to relieve stomach ache by neutralizing acid in the stomach. What experiments would you carry out:
(a) to show that the tablets do in fact neutralize acids.
(b) to discover which tablet neutralizes more acid?

16 Design a poster illustrating the manufacture and uses of oxygen.

Iron doesn't usually burn in air

8 Competition among the elements

8.1 Introduction

Different elements burn in oxygen with different vigour. Some elements, such as sodium and magnesium, react extremely rapidly, burning with bright flames. Others, such as iron and copper, react very slowly and do not burn with flames.

As a result of these reactions we arranged these four metals in a rough order of reactivity called the activity series. We put the most vigorous metals, sodium and magnesium, at the top of the series and the least reactive (copper) at the bottom. How can we test our idea of an activity series?

Try to answer these questions

1 Which element, sodium or copper, has the greater liking for oxygen? Which element is the keener to combine with oxygen?

2 Suppose these two metals are placed separately in water. Which one is more likely to remove oxygen from water to form its oxide?

3 If oxygen is removed from water, what other substance is also produced? (Remember, water is hydrogen oxide).

8.2 The reactions of metals with water

SODIUM

A small piece of sodium placed on cold water reacts very vigorously, melting due to the heat produced. The molten bead of sodium moves around the surface of the water fizzing and becoming smaller as it reacts (Figure 8.1). **You must not try this reaction yourself.**

If universal indicator is added to the water, it changes colour during the reaction from yellow to blue (pH about 10) showing that the solution is becoming alkaline. The fizzing noise suggests that a gas is being produced during the reaction. This gas can be collected by holding the sodium below the surface of the water inside a steel mesh. Bubbles of a colourless gas rise from the sodium and can be collected in an inverted test tube full of water (Figure 8.2).

The gas explodes with a 'pop' when a lighted taper is placed in the mouth of the tube, showing the gas to be hydrogen. It seems that sodium has such a strong liking for oxygen that it can break the link between hydrogen and oxygen in water.

Figure 8.1

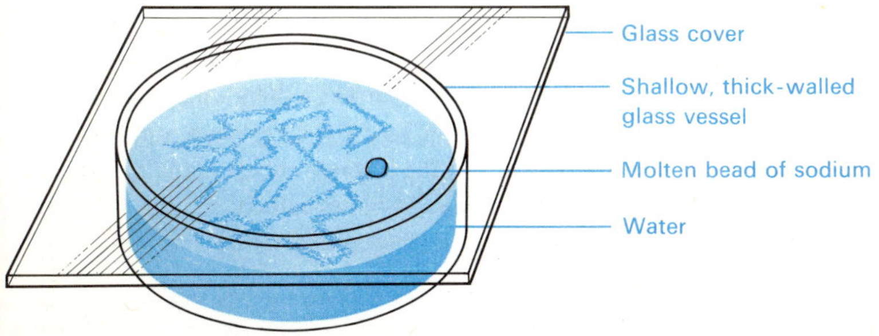

Figure 8.2

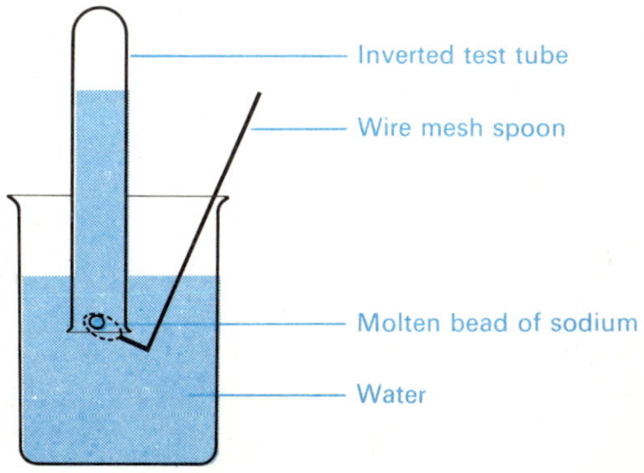

Figure 8.3

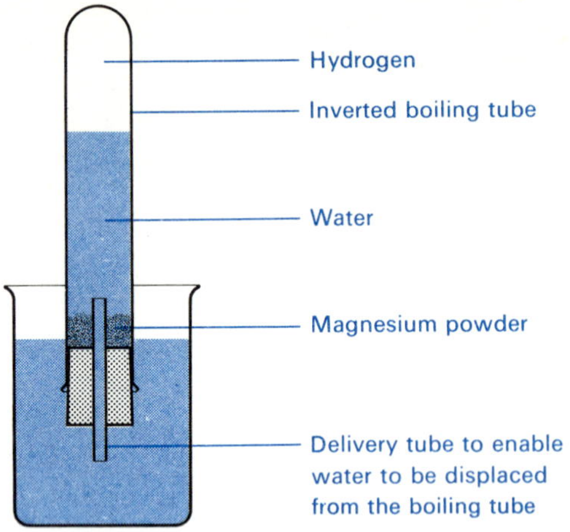

Figure 8.4

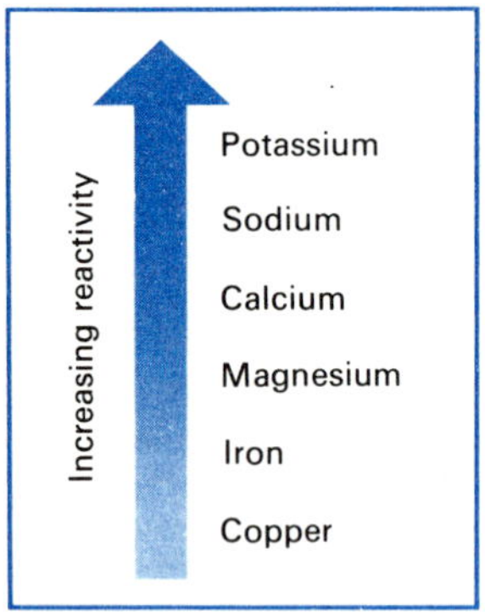

Figure 8.5

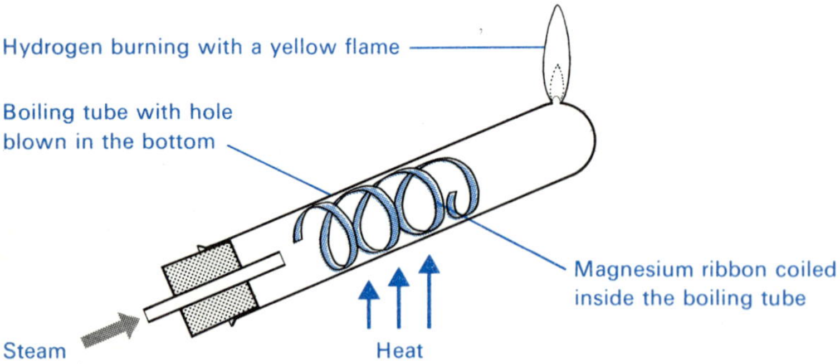

A competition takes place between hydrogen and sodium for oxygen. The sodium wins, forming sodium oxide, leaving hydrogen behind. The equation for this reaction is:

$$\text{sodium}_{(s)} + \text{water}_{(l)} \longrightarrow \text{sodium oxide}_{(s)} + \text{hydrogen}_{(g)}$$
(*hydrogen oxide*)

The sodium is keener than the hydrogen to combine with oxygen. Pure sodium oxide is a solid at room temperature, but no solid remains at the end of the experiment. This is because the sodium oxide reacts with water to form an alkaline solution of sodium hydroxide:

$$\text{sodium oxide}_{(s)} + \text{water}_{(l)} \longrightarrow \text{sodium hydroxide}_{(aq)}$$

POTASSIUM

When a small piece of potassium is placed on water a similar reaction to sodium occurs. However, the reaction is even more violent and a lilac flame is seen. The colour of the flame is due to burning potassium. **You must not try this experiment yourself.**

The reaction shows that potassium is an even better competitor for oxygen than sodium since it removes oxygen from water more rapidly and more violently. Potassium oxide and hydrogen are formed. The potassium oxide then reacts with water to form potassium hydroxide which dissolves in the water to form an alkaline solution:

$$\text{potassium}_{(s)} + \text{water}_{(l)} \longrightarrow \text{potassium oxide}_{(s)} + \text{hydrogen}_{(g)}$$

$$\text{potassium oxide}_{(s)} + \text{water}_{(l)} \longrightarrow \text{potassium hydroxide}_{(aq)}$$

MAGNESIUM

If a short strip of magnesium ribbon is placed in cold water, no reaction appears to take place. However, when magnesium powder is placed in an inverted boiling tube filled with water a slow reaction takes place and hydrogen collects in the inverted tube (Figure 8.3). It takes about 24 hours before the tube contains sufficient hydrogen to give the 'pop' test:

$$\text{magnesium}_{(s)} + \text{water}_{(l)} \longrightarrow \text{magnesium oxide}_{(s)} + \text{hydrogen}_{(g)}$$

The solution becomes alkaline because

magnesium hydroxide solution is formed:

magnesium oxide$_{(s)}$ + water$_{(l)}$ ⟶ magnesium hydroxide$_{(aq)}$

Magnesium hydroxide is only slightly soluble in water: solid white magnesium hydroxide forms when the solution becomes saturated. Magnesium is obviously a poorer competitor for oxygen than either sodium or potassium; the reaction is very much slower and less vigorous. On this evidence magnesium would be placed after potassium and sodium in the activity series.

CALCIUM

Calcium reacts with water more rapidly than magnesium but not so vigorously as sodium or potassium. The calcium sinks to the bottom and bubbles of hydrogen are produced. A test tube of gas can be collected in a few minutes and the solution becomes alkaline. The calcium hydroxide which forms is not very soluble in water andso, after a time, calcium hydroxide forms as a white solid.

IRON

Iron filings rust very slowly when left in cold water for several days provided oxygen is also present. Hydrated brown iron oxide forms which is insoluble in water.

COPPER

Copper does not react with water to form copper oxide and hydrogen, no matter how long the two substances are left in contact.

8.3 The activity series

We can now extend our activity series to include six metals: potassium, sodium, calcium, magnesium, iron and copper (Figure 8.4).

The order of reactivity of metals with water is very similar to the order of reactivity with oxygen (Figure 7.5, page 89). This is not really very surprising since in each case the metal is reacting to form its oxide. The two processes can be summarized as:

metal$_{(s)}$ + oxygen$_{(g)}$ ⟶ metal oxide$_{(s)}$

metal$_{(s)}$ + hydrogen oxide$_{(l)}$ ⟶ metal oxide$_{(s)}$ + hydrogen$_{(g)}$
(*water*)

It is important, though, to remember that the activity series is only a guide to the relative reactivities of the metals. The state of the metal (whether it is in the form of lumps or in a fine powder) and other conditions, such as temperature, may bring about great variations in reactivity. For example, cold iron filings do not react with chlorine, but hot filings immediately produce thick brown fumes of iron chloride. Thus, the activity series built up from one series of similar reactions may not always coincide exactly with an activity series based on another series of reactions.

Try to answer these questions

1 Can we place hydrogen in the activity series?
2 Which metals in our series can remove oxygen from water?
3 Which metals in our series have a greater liking than hydrogen for oxygen?
4 Which metals in our series cannot remove oxygen from water?
5 Where would you place hydrogen in the series in terms of its liking for oxygen?

8.4 The reaction of metals with steam

How would you expect metals to react with steam? The steam is hotter than water and as a result of this the higher temperature makes the reaction faster. Magnesium reacts slowly with water but vigorously with steam (Figure 8.5).

First, steam is passed through the boiling tube to remove all the air. The magnesium is then heated strongly. After a while, the magnesium reacts with the steam giving a bright white flame (similar to the flame when magnesium burns in oxygen). The gas leaving the tube is hydrogen, which burns with a yellow flame. A white solid, magnesium oxide, remains in the tube after the reaction has finished:

magnesium$_{(s)}$ + steam$_{(g)}$ ⟶ magnesium oxide$_{(s)}$ + hydrogen$_{(g)}$
(*hydrogen oxide*)

Sometimes, the magnesium gets so hot that it reacts with the glass of the boiling tube to form a black substance; but the white oxide can still be seen on the inside of the tube.

Heat will speed up most chemical reactions.

However, copper, the lowest metal in our activity series, does not react with steam, even at high temperatures. Copper and lead can be used as hot-water pipes in houses because they do not react with water or steam. The vigour of the reactions of magnesium, iron and copper with steam are in the order that we might have predicted from our activity series. This is further evidence that our order of reactivity is correct.

Another way of comparing the activity of metals is to see how they compete with each other for oxygen.

Molten iron from the thermit reaction runs into a mould round the two rails to be joined, melting the ends. When the iron has cooled, the mould is broken open, and the excess metal trimmed off

Try to answer these questions

1 How would you expect copper oxide and magnesium to react when heated?

2 Which metal, magnesium or copper, has the greater desire to combine with oxygen?

3 Will the magnesium remove the oxygen from copper oxide to form magnesium oxide and copper?

4 How would you expect magnesium oxide and copper to react?

5 Will copper remove oxygen from magnesium oxide to form copper oxide and magnesium?

The competition of metals for oxygen is used in the thermit or Goldschmidt process to prepare manganese, chromium and titanium from their oxides. The oxides of these metals are mixed with powdered aluminium or another metal high in the activity series and the mixture is ignited with a fuse of burning magnesium.

Aluminium has a greater desire for oxygen than manganese, chromium or titanium, and a vigorous exothermic reaction takes place, forming aluminium oxide and the metal:

$$\text{aluminium}_{(s)} + \text{chromium oxide}_{(s)} \xrightarrow{\textit{heat}} \text{aluminium oxide}_{(s)} + \text{chromium}_{(s)}$$

A mixture of powdered aluminium and iron oxide (known as 'thermit') is used in incendiary bombs and for welding. The aluminium is keener than iron to combine with oxygen. It removes oxygen from iron oxide to form aluminium oxide and iron. The reaction is very exothermic and the iron is produced in a molten state. As the molten iron solidifies it will join (weld) two pieces of steel together. Continuous welding on Inter-City railway lines has extended the average 'life' of the track from 23 to 30 years.

8.5 Redox

Many reactions which occur in nature involve substances combining with oxygen to form oxides. Burning, respiration and rusting are three important examples. We say that substances are **oxidized** when they combine with oxygen, and we call the process **oxidation.**

But, if one substance combines with oxygen another substance (possibly oxygen itself) must

Figure 8.6

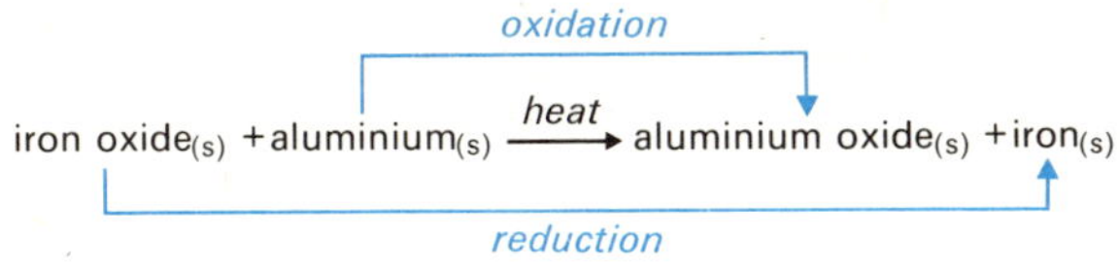

Figure 8.7

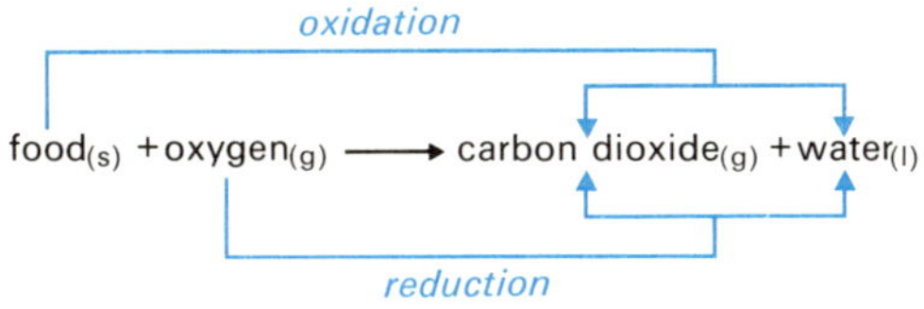

A small hand spectroscope

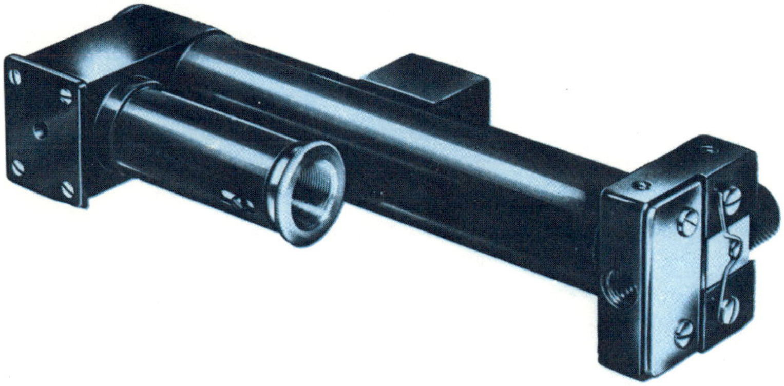

Figure 8.8

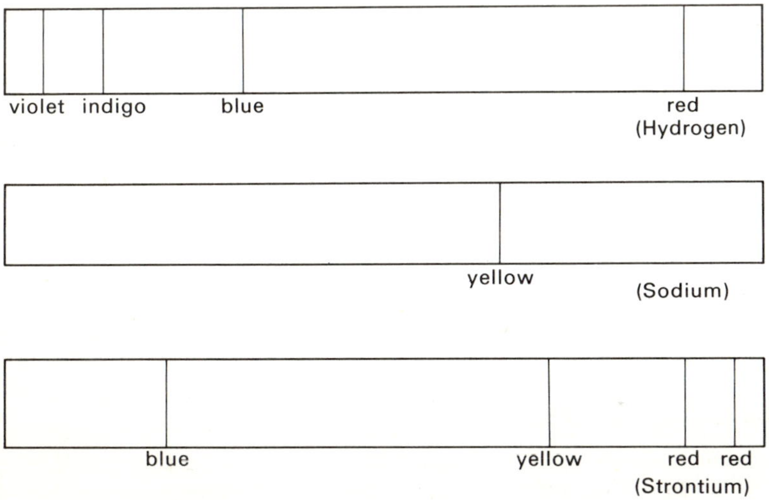

lose oxygen. We say that substances which lose oxygen in chemical reactions are **reduced** and we call the process **reduction.**

Oxidation and reduction must always occur together, for if one substance combines with oxygen (and is oxidized) another substance must lose oxygen (and be reduced). We call the combined process **redox** (**red**uction + **ox**idation). Figure 8.6 shows what happens when iron oxide reacts with aluminium.

Aluminium is oxidized, combining with oxygen to form aluminium oxide. Iron oxide is reduced, losing its oxygen to form iron. The whole process involves reduction and oxidation (redox).

The food we eat is oxidized to carbon dioxide and water. The oxygen that we breathe is reduced in the same process to carbon dioxide and water (Figure 8.7).

Try to answer these questions

Which substance is oxidized and which is reduced in each of the redox reactions below?

1 sodium + water ⟶ sodium oxide + hydrogen

2 hydrogen + oxygen ⟶ water

8.6 Hydrogen

Hydrogen is the lightest of all elements. The name 'hydrogen' means 'water producer'. Although hydrogen is only the ninth most abundant element in the Earth's crust it is by far the most abundant element in the Universe. Recently, scientists have estimated that about 80% of the weight of the Universe is hydrogen. The sun and many other stars contain many billions of tonnes of hydrogen which are being continuously transformed into helium.

Elements heated enough become incandescent (glow): magnesium becomes incandescent when it burns in oxygen. Incandescent materials give off (emit) light, which can be split up into different colours (called a spectrum) in a spectroscope. Each element produces its own characteristic spectrum which differs from that of all other elements.

Nowadays, scientists can identify elements from the different colours of light which they emit as part of their spectra. Figure 8.8 shows the visible part of the spectra of light emitted by hydrogen, sodium and strontium. Each spectrum is composed of a series of sharp lines which correspond to particular colours of light. The four visible lines in the spectrum emitted by hydrogen are red, blue, indigo and violet. The one line in the spectrum for sodium is yellow.

Hydrogen does not occur in the air, but it occurs in many compounds, including water. Our bodies also contain hydrogen in important biological substances such as fats, proteins, carbohydrates and nucleic acids. Hydrogen accounts for about 10% of your weight.

8.7 What is hydrogen used for?

In the Haber process, hydrogen combines with nitrogen to form ammonia (hydrogen nitride). The process occurs under great pressure and in the presence of a brown iron oxide catalyst.

$$\text{hydrogen}_{(g)} + \text{nitrogen}_{(g)} \longrightarrow \underset{\textit{(hydrogen nitride)}}{\text{ammonia}_{(g)}}$$

Ammonia is an important starting material in producing ammonium compounds and nitric acid.

The manufacture of margarine and other fats involves the treatment of vegetable oils (palm oil, coconut oil and peanut oil) with hydrogen at high pressures and in the presence of a nickel catalyst. Inedible vegetable oils can be converted by a similar process to solid fats which are used to manufacture soap.

Liquid hydrogen is used as a fuel in rockets. The hydrogen burns in oxygen to liberate the energy required to propel the later stages of the rocket.

8.8 How do we get hydrogen?

IN INDUSTRY

Hydrogen is formed when metals high up in the activity series react with water. For example:

$$\text{sodium}_{(s)} + \text{water}_{(l)} \longrightarrow \text{sodium oxide}_{(s)} + \text{hydrogen}_{(g)}$$

Metals in the middle of the activity series (such as iron) will react with steam to produce a steady supply of hydrogen:

$$\text{iron}_{(s)} + \text{steam}_{(g)} \longrightarrow \text{iron oxide}_{(s)} + \text{hydrogen}_{(g)}$$

Although hydrogen has been produced commercially in the past by passing steam over red-hot iron, the method is not really suitable for preparing large quantities of hydrogen. So today, hydrogen is produced in one of three main ways:

1 By reducing steam to hydrogen with red-hot coke (carbon) at about 1000°C:

$$\underset{\textit{(carbon)}}{\text{coke}_{(s)}} + \underset{\textit{(hydrogen oxide)}}{\text{steam}_{(g)}} \xrightarrow{\textit{heat}} \text{carbon monoxide}_{(g)} + \text{hydrogen}_{(g)}$$

Carbon, like some metals, can remove oxygen from steam at high temperatures. The product, a mixture of carbon monoxide and hydrogen (called water gas), is a very good fuel since both these gases burn in oxygen exothermically.

2 By electrolysing a concentrated solution of sodium chloride. Chlorine is produced at the anode while hydrogen and a solution of caustic soda (sodium hydroxide) are produced in the cathode processes. (Section 5.7, page 60). In some countries where electrical power is cheap hydrogen is prepared by the electrolysis of dilute sulphuric acid.

3 By reacting crude oil fractions or natural gas with steam in the presence of catalysts at high temperatures and pressure.

Hydrogen is a valuable product of oil refineries.

IN THE LABORATORY

Small quantities of hydrogen can be prepared in the laboratory by the reaction of zinc with dilute hydrochloric acid or dilute sulphuric acid.

8.9 Can hydrogen reduce metal oxides?

Try to answer these questions

1 Which metals have a much greater attraction than hydrogen for oxygen? Which metals can remove oxygen from water vigorously?

2 Which metals have a weaker attraction than hydrogen for oxygen?

3 Which of the following four metal oxides will be reduced to the metal when hydrogen is passed over the heated oxide?

Sodium oxide
Iron oxide
Magnesium oxide
Copper oxide

Here is how we can find out which metal oxides can be reduced by hydrogen.

Pass a stream of hydrogen over the heated metal oxide using the apparatus in Figure 8.9. Allow plenty of time for the air to be displaced from the silica tube before you light the excess of hydrogen at the jet.

Copper oxide and lead oxide are reduced to copper and lead easily. Black copper oxide changes to red-brown copper, and yellow lead oxide to dark-grey lead. We can test whether the oxide has been reduced to a metal by seeing if the product will conduct electricity:

$$\underset{(black)}{\text{copper oxide}_{(s)}} + \text{hydrogen}_{(g)} \longrightarrow \underset{(red\ brown)}{\text{copper}_{(s)}} + \underset{(hydrogen\ oxide)}{\text{water}_{(g)}}$$

$$\underset{(yellow)}{\text{lead oxide}_{(s)}} + \text{hydrogen}_{(g)} \xrightarrow{heat} \underset{(dark\ grey)}{\text{lead}_{(s)}} + \underset{(hydrogen\ oxide)}{\text{water}_{(g)}}$$

Copper and lead have a weaker attraction for oxygen than hydrogen (copper does not remove oxygen from water or steam). Hydrogen can, therefore, remove oxygen from copper oxide and lead oxide to form copper and lead, respectively. In fact, hydrogen is used to manufacture some metals from their oxide ores.

Neither iron nor zinc can be produced from their oxides in the apparatus in Figure 8.9, but hydrogen will reduce these oxides at higher temperatures. In fact, methods have been developed for manufacturing iron by reducing iron oxide by hydrogen.

Metals, such as iron and zinc, a little higher in the activity series than copper and lead, have a slightly greater attraction than hydrogen for oxygen. Iron, for example, will remove oxygen from steam to form iron oxide and hydrogen. Thus, it is difficult to remove oxygen from iron oxide.

Sodium oxide, calcium oxide and magnesium oxide are extremely difficult to reduce to the metal. They cannot be reduced by hydrogen even at high temperatures.

Again, this is what we might have predicted from the activity series. Sodium, calcium and magnesium are all very keen to react with oxygen and form their oxides. They all have a much greater attraction than hydrogen for oxygen. Consequently, they can easily stick to their oxygen in competition with hydrogen, so their metal oxides will not be reduced.

The results of these experiments confirm the patterns of reactivity which we saw previously between the metals and oxygen and between the metals and water.

8.10 Manufacturing metals

Many important metals exist in the Earth's crust as oxides or as substances which can be converted to oxides when heated in air (e.g., sulphides):

$$\text{metal sulphide}_{(s)} + \underset{(in\ air)}{\text{oxygen}_{(g)}} \xrightarrow{heat} \text{metal oxide}_{(s)} + \text{sulphur dioxide}_{(g)}$$

Iron occurs in the Earth's crust as iron ore (brown iron oxide). Copper, zinc and lead usually occur as their sulphides (copper pyrites, zinc blende and galena), which are converted to oxides when heated in air.

All these metals are produced industrially by reducing the oxide to the metal. One method of removing the oxygen from these oxides is to

Figure 8.9

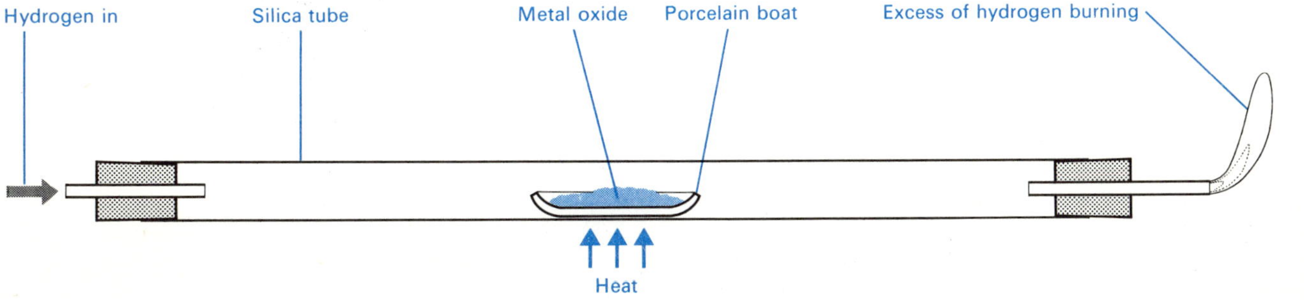

react the oxide with a metal which has a greater tendency to combine with oxygen. This would need reactive metals, which are very expensive. Therefore, some other means of removing oxygen must be used.

Hydrogen gas has been used to reduce the oxides of metals low in the activity series such as copper and lead:

$$\text{copper oxide}_{(s)} + \text{hydrogen}_{(g)} \xrightarrow{\textit{heat}} \text{copper}_{(s)} + \underset{(\textit{water})}{\text{hydrogen oxide}_{(g)}}$$

However, metals in the middle of the activity series, such as iron and zinc, have a stronger attachment than hydrogen to oxygen and their oxides cannot be reduced easily with hydrogen.

During the Iron Age our ancestors discovered that iron ore could be reduced to iron by heating with charcoal (carbon). Carbon is still used in the large scale manufacture of iron from iron ore. Large quantities of carbon are available nowadays in the form of coke. The coke and iron are heated to about 1500°C in large blast furnaces which hold up to 2000 tonnes of material. The iron oxide is reduced to molten iron which trickles to the bottom of the furnace and can be tapped off.

Not all metal oxides are reduced by carbon. The active metals, sodium, potassium and magnesium, have a greater desire than carbon or hydrogen for oxygen and so their oxides are not readily reduced. These metals liberate a great deal of energy when they combine with oxygen. We have to put energy back into the compound to split it into its elements. Chemists can do this by electrolysis.

Most of the active metals, such as sodium, potassium and magnesium, are manufactured by electrolysing their molten chlorides (section 5.7, page 59).

Metals which can be obtained easily from their ores such as copper, silver and lead have been made since 3500 BC. Iron, which required charcoal to reduce the iron ore, has been made since about 1000 BC. The more reactive metals whose oxides cannot be reduced by carbon or hydrogen, and which have to be produced by electrolysis of molten compounds have only been extracted in the last 170 years.

A blast furnace like this can produce up to 2000 tonnes of iron in 24 hours

This iron sword dates from the 1st century BC: it is more than 300 years younger, but much more corroded, than the gold dagger handle on page 42

Summary

1 The order of reactivity of metals with water is similar to the order of reactivity of metals with oxygen. In each case the metal is converted to its oxide:

$$\text{metal}_{(s)} + \text{oxygen}_{(g)} \longrightarrow \text{metal oxide}_{(s)}$$

$$\text{metal}_{(s)} + \text{water}_{(l)} \longrightarrow$$
(*hydrogen oxide*)
$$\text{metal oxide}_{(s)} + \text{hydrogen}_{(g)}$$

2 Competition of the metals with hydrogen and between each other for oxygen confirms the order of metals in the activity series.
3 Using the activity series a pattern of reactivity emerges which summarizes our knowledge and makes it possible to predict the likelihood and vigour of other reactions.
4 When substances combine with oxygen they are oxidized and the process is called oxidation. When substances lose the oxygen they contain they are reduced and the process is called reduction. Reduction and oxidation occur together (one substance losing oxygen, another substance gaining oxygen). The combined process is called redox.
5 Hydrogen does not occur in the air, but it does occur in large quantities combined with other substances (e.g., in water, in crude oil and in biological chemicals such as fats, proteins and carbohydrates).
6 The main methods of obtaining hydrogen industrially are:
(a) by reducing steam to hydrogen with red-hot coke.
(b) by electrolysing concentrated sodium chloride solution.
(c) by catalysing the reaction between crude oil fractions and steam.
7 Many important metals (copper, iron, lead and zinc) exist in the Earth's crust as oxides or as substances (e.g., sulphides) which can be converted to oxides when heated in air. The oxides can be reduced to the metal with carbon (coke).
8 The more reactive metals (potassium, sodium, calcium and magnesium) whose oxides cannot be reduced by carbon or hydrogen are obtained by electrolysis of their molten compounds. The metals are liberated at the cathode during electrolysis.

Study questions

1 Give three pieces of evidence to show that iron is a more reactive metal than copper. Write word equations for the reactions you discuss.
2 (a) Where would you place gold in the activity series?
(b) How would you expect gold to react with the following?
Water Oxygen
Steam
3 Hydrogen is usually prepared in the laboratory by the reaction of dilute hydrochloric acid on zinc in a Kipp's apparatus (Figure 8.10).
(a) Write a word equation for the reaction of zinc with dilute hydrochloric acid.
(b) Why is zinc chosen?
(c) Why is sodium not used in place of zinc?
(d) Why is copper not used in place of zinc?
(e) Could sulphuric acid be used in place of hydrochloric acid?
(f) Explain how the Kipp's apparatus works. Start your explanation as follows; 'When tap, *T*, is opened . . .' (The three chambers of the apparatus have been labelled *A*, *B* and *C*.)
(g) What are the advantages of using a Kipp's apparatus to prepare gases which are needed from time to time?

Figure 8.10

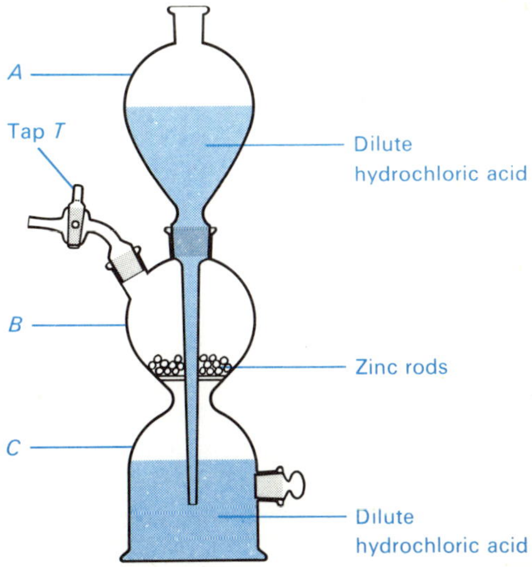

4 Draw up a table to show the differences in the properties of water, hydrogen and oxygen.

5 Consider the following metal oxides:

Magnesium oxide	Sodium oxide
Copper oxide	Red mercury oxide

(a) Which are basic?
(b) Which are alkaline?
(c) Which give a green or blue colour with universal indicator?
(d) Which can be reduced to the metal by heating with a bunsen?
(e) Which can be reduced to the metal by heating in hydrogen gas?
(f) Which can be reduced to the metal by heating with powdered iron?

6 Three elements were burned in oxygen and the products were dissolved in water and universal indicator added. Two of the final solutions were blue and the other was red. The elements could have been (state true or false for each answer):
(a) sulphur, calcium and magnesium.
(b) sulphur, carbon and magnesium.
(c) potassium, sodium and phosphorus.
(d) carbon, hydrogen and sodium.
(e) sulphur, sodium and iron.

7 Red-hot carbon will remove oxygen from both copper oxide and zinc oxide but not from magnesium oxide. Zinc will remove oxygen from copper oxide.
(a) Which of these metals has the greatest desire to react with oxygen?
(b) Which of these metals has the weakest desire to react with oxygen?
(c) Write the three metals in order of activity, putting the most reactive metal first.
(d) Where would you place carbon in this activity series?
(e) How vigorously would you expect zinc to react with oxygen?
(f) How vigorously would you expect zinc to react with water?
(g) How would you expect heated magnesium to react with carbon dioxide?

8 Steam is passed over red-hot coke and a mixture of two gases (*A* and *B*) is produced. This mixture is then burned in pure oxygen and, when the products of the reaction are cooled to room temperature, a colourless liquid, *C*, and a colourless gas, *D*, remain. *D* reacts with a dilute solution of calcium hydroxide (lime water) to form a white precipitate, *E*. *D* allows magnesium to burn in it forming two solids, *F* and *G*. When *D* is passed over red-hot coke the gas, *A*, is formed.
(a) Identify the substances *A*–*G*.
(b) Write word equations for all the reactions involved.

9 In the following chemical changes, is the first substance being reduced?
(a) Lead oxide to lead.
(b) Lead oxide to lead dioxide.
(c) Carbon dioxide to carbon monoxide.
(d) Carbon monoxide to carbon.
(e) Water to hydrogen.
(f) Copper to copper oxide.

10 (a) Write word equations for the reactions which take place when the following react with oxygen:

Sodium	Magnesium
Carbon	Sulphur

(b) What two groups can the products be placed in?
(c) How does the vigour of their reaction with oxygen relate to their position in the periodic table?

11 Some rust was dried carefully for several days in a desiccator. It was then heated strongly in a vacuum, and water condensed on the cooler parts of the test tube. The solid residue was heated strongly in a stream of hydrogen. Water condensed on the cooler parts of the apparatus and a dark grey magnetic solid remained which conducted electricity.
(a) Why was the rust dried so carefully?
(b) Where did the water come from when the rust was heated in a vacuum?
(c) Write a word equation for the process taking place when rust is heated in a vacuum.
(d) How is the water produced when the residue is heated in a stream of hydrogen?
(e) What do you think the residue is? How does the residue differ from the initial rust?
(f) Describe what you think rust is.

Index

acetic acid, 57
acetone, 19, 25
acetylene, 83, 94
acids, 56 – 58, 88 – 90
  conduction by, 56 – 58
  definition, 89
acidic oxide, 89
activity series, 88, 101, 102
air, 8, 11, 42, 48, 65 – 76
  and burning, 65 – 72
  exhaled air, 73
  fractional distillation, 84
  gases in, 72
  % oxygen in, 70
alcohol, 21, 24, 25
alkalis, 89 – 94, 99 – 101
alloys, 8, 48
  to prevent rusting, 76
aluminium, 8, 48, 55
  manufacture, 59, 60
  in thermit reaction, 102
aluminium oxide, 59, 60, 102
amino acids, 24
ammonia, 8, 24, 86
  manufacture, 104
ammonium nitrate, 94
ammonium sulphate, 91, 94
anhydrous, 33
anode, 56
  products at, 58, 59
aqueous, 32
argon, 8, 72
asbestos, 7
atmosphere (see *air*)
atmospheric pressure, 26
atomic bomb, 70
'azote', 71

bacteria, 48
  in tooth decay, 92, 93
baking powder, 34
basic oxides, 89
battery, 44
benzene, 56
bleach, 60
bleaching powder, 94
blood, 25, 74
blotting paper, 24
blast furnace, 106
boil, 19, 20, 26, 31, 33
boiling point, 20, 21, 26, 43
bomb (atomic), 70
bond (chemical), 46, 47
bread, 30, 31, 33
breathing (see *respiration*)
  and burning, 74
  gas exchange during, 73, 74
  respiration and redox, 102
brine (see *sodium chloride*)
bromine, 42, 58
bronze, 83
burning, 65 – 72
  and breathing, 74
  of a candle, 66
  early investigations, 71
  importance, 67
  property of, 10
  and redox, 102
  weight changes on, 69
butanol (n-), 24

cables, 47, 55
calcium, 49, 59, 88, 101
calcium carbide, 94
calcium carbonate, 48
calcium hydroxide, 34, 101
  to cure acid soils, 91, 92
  lime water, 56
  uses, 94
calcium oxide, 91, 105
calcium phosphate, 49
candle (burning of), 66, 67
carbohydrates, 49, 74, 86, 104
carbon, 31, 35, 41 – 44, 47, 49, 104
  reaction with oxygen, 88
carbon dioxide, 32, 35, 88
  % in air, 73
  test for, 34
carbon disulphide, 45
carbon monoxide, 68, 104
carbonic acid, 89
catalase, 86
catalysis, 86
catalysts
  in the Haber process, 104
  manganese dioxide, 85, 86
cathode
  definition, 56
  products at, 58, 59
caustic soda (see *sodium hydroxide*)
cellulose acetate, 55, 94
centrifuge, 17, 18
chalk, 47, 48
change of state, 31
charcoal (see also *carbon*), 35, 106
chemical change (see also *chemical reaction*), 31
chemical reaction, 31, 45, 46, 92
  effect of temperature on, 101
chlorides, 88, 106
chlorine, 8, 48
  manufacture, 59
chromatography, 23 – 25, 41
chromium, 48, 60, 76
  manufacture, 102
classification
  elements, 42
  metals and non-metals, 43, 59, 90
  mixtures and compounds, 46
  pure substances and mixtures, 11
  separation processes, 25
  solids, liquids and gases, 10
clay, 48
Clean Air Act, 68
coal, 8, 31, 67
coal tar, 8
coke (see also *carbon*), 7, 104, 106
collection of gases, 34
  after electrolysis, 58
  over water, 35
colour(s), 10, 23, 25, 31, 47
  and indicators, 88 – 91

combine(d), 44, 45
combustion (see *burning*)
component(s), 11, 21, 46, 47
composition, 46, 47
compound, 44–47
concentrated, 19
condensation, 19
condenser, 20
conduction of electricity
  by acids, 56, 57
  by compounds, 55
  by elements, 43
  by metal / non-metal compounds, 56–58
  by metals, 56, 57
  by non-metal / non-metal compounds, 56, 57
conductors
  compounds, 55
  elements, 43, 44
conservation of mass (law of), 69
constant composition (law of), 46
constituent(s), 45
control experiment, 75
cooking, 65, 67
copper, 42, 44–46, 58, 101, 106
change in weight on heating, 34, 36
  in electroplating, 60, 74
  uses, 47, 55
copper oxide, 36, 44–46
  reduction of, 105
copper sulphate
  decomposition, 33, 34
  electrolysis, 56, 58
  use, 94
copper pyrites, 105
cotton, 7, 94
crystals, 19
crystallization, 19
crude oil, 8, 15, 20, 104
cyanide, 90

decantation, 15, 22
decay (tooth), 92, 93
decomposition, 33, 34, 41
density, 10, 23, 42, 43
dew formation, 73
diamond, 47
dilute, 19
dissolve, 15, 20
distillate, 20
distillation, 19–22, 41
  fractional, 20, 21
distilled water, 20, 26
drugs, 7, 25, 94
'dry ice', 32
dyes, 24, 25, 94

dynamite, 94

Earth, 7, 48
Earth's crust, 42
  % elements in, 48, 49
  metal ores in, 105
electrical decomposition, 41
electrical energy, 56, 59
electricity
  conduction of (see *conduction of electricity*)
  current for electroplating, 60
  generation of, 65
electrodes, 56, 57
  products at, 59
electromagnetic separation, 23
electrolysis
  of compounds, 56
  definition, 41
  industrial applications, 59
  to manufacture hydrogen, 104
  to manufacture metals, 106
electrolyte, 56, 58, 59
electroplating, 59, 60, 76
  conditions for, 60
  metals used in, 59
  with silver, 60
elements, 41, 42, 44, 45, 48, 49
  competition among, 99–106
  extraction, 59, 60
  reaction with oxygen, 87–90
energy
  electrical, 56, 59
  nuclear, 70
  thermal, 31, 34, 41, 74, 86
enzymes, 86
equations, 31, 32
ethanol, 24, 56, 57
evaporation, 11, 19, 20
exhaled air, 73, 74
exothermic reaction, 45, 46, 67, 102
explosives, 67, 68, 94
extraction of elements, 59, 60

fat(s), 22, 49, 74, 86, 94, 104
fertilizers, 8, 9, 91, 94
fibres, 7, 8
filter bed, 16
filter funnel, 16
filter paper, 16, 18, 24
filtrate, 16, 23
filtration, 16–18, 23, 41
flame(s), 67
floatation, 22, 23
fog, 72, 73
food(s), 31
  and combustion, 74
  and cooking, 65
  enzyme breakdown of, 86
  and redox, 103
  and tooth decay, 93
fractional distillation, 20, 21
  of air, 84
freezing point, 26
frost (formation of), 73
fuels, 65, 67

gain in weight, 36, 44
galena, 105
galvanizing, 76, 94
gases, 10, 11, 19, 32, 42
  collection of, 34, 35
germanium, 48, 55
gold, 8, 42, 47, 60, 88
Goldschmidt process, 102
graphite (see also *carbon*), 43, 44
gravity, 17

Haber process, 104
haematite (see *iron ore*)
heat, action on materials, 31–36, 41
heat energy, 74
helium, 72, 103
human body (% elements in), 49
hydrates, 76
hydrochloric acid
  in preparation of hydrogen, 104
  uses, 94
hydroelectricity, 60
hydrogen, 41, 45, 46, 48, 49, 103–105
  explosion with oxygen, 67
  lab. preparation, 104
  manufacture, 104
  from metals with water, 99–101
  production in electrolysis, 58–60
  reduction of metal oxides, 104, 105
  rocket fuel, 83
  test for, 42, 67
  uses, 104
hydrogen oxide (see also *water*), 45
hydrogen peroxide, 85, 86

immiscible, 22
incandescence, 67
increase in weight, 36, 44
indicators, 88–90
indigestion cures, 92
inert gases (see *noble gases*)
ink, 23, 24
insoluble, 18, 19, 22, 23
insulators, 55
iodine, 32, 33
iron, 7, 23, 36, 42, 45, 46, 48, 55, 98
  manufacture, 106
  in manufacture of hydrogen, 104
  reaction with oxygen, 88

reaction with water, 101
rusting of, 74–76
iron chloride, 88
iron ore, 7, 49, 106
iron oxide (see also *iron ore*), 101
formation in rusting, 76
Haber process catalyst, 104
in thermit reaction, 102
iron sulphide, 45, 46, 88
irreversible reaction, 33

jewellery, 47, 60, 61

kerosene (as rocket fuel), 83
khroma, 25
krypton, 72

Lavoisier, 71, 72
laws (chemical), 69
conservation of mass, 69, 70
constant composition, 46
lead, 42, 47, 48, 58, 105, 106
lead bromide (electrolysis of), 56, 58, 59
lead oxide, 35, 105
lime, 7
limestone, 48, 49
lime water, 34, 56
liquids, 10, 32, 42
litmus, 90
liver (catalase in), 86
loss in weight, 34

magnesium, 8, 33, 34, 36, 42, 44, 59, 100, 101
manufacture, 106
reaction with oxygen, 88
in thermit reaction, 102
magnesium hydroxide, 101
magnesium oxide, 36, 44, 86, 100, 101, 105
magnetic separation, 23, 45
malleable, 42, 47
man-made, 7, 8, 42
manganese, 102
manganese dioxide, 85, 86
mass, 69
mass spectrometer, 42
melting, 31–33
melting point, 26, 43
mercury, 35, 41, 42, 44, 56, 71, 85
mercury oxide, 33–35, 41, 71, 85
metal / non-metal compounds, 56–59
metals, 7, 8, 42–44, 47
conduction of, 59
difference in reaction to non-metals, 90
manufacture, 105, 106
reaction with steam, 101, 102
reaction with water, 99–101

milk, 15–17, 22, 23, 36, 90
mineral ore, 8
miscible, 20, 21
mist, 16, 66
mixtures, 11, 15, 23, 25, 45–47, 91
mud, 11, 15–18, 23

naphthalene, 56
natural gas, 67
naturally occurring, 7, 42, 48
neon, 8, 72
neutral, 88, 92
neutralization, 91, 92
nickel, 48, 55, 60, 104
nitric acid, 8, 86, 90, 94, 104
nitrochalk, 94
nitrogen, 8, 49, 71, 84
noble gases, 71, 72, 85
non-metal / non-metal compounds, 56–59
non-metals, 42–44, 47, 90
North Sea gas, 67
nuclear reactions, 42, 70
Nylon, 7

octane, 11
oil(s), 19, 22, 36
as a fuel, 67
in margarine manufacture, 104
in soap making, 94
ores, 22, 42, 105
oxidation, 102, 103
oxides, 36, 44
acidic, 89
alkaline, 89
basic, 89
of metals, 88
of non-metals, 88
ores, 105
and redox, 102
reduction of, 104, 105
oxy-acetylene, 8, 83
oxygen, 8, 35, 36, 41, 45, 46, 48, 49, 83–90
% in air, 70
in burning, 72
lab. preparation, 85
manufacture, 59, 60, 83, 84
production in electrolysis, 58–60
reaction with elements, 87–90
in rusting, 75, 76
test for, 35
uses, 83

paraffin, 20
patterns in reaction, 43
of metals and non-metals, 90

in reduction of metal oxides, 105
permanent changes, 31, 33
petrol, 11, 15, 20
pH
changes in mouth, 93
changes in neutralization, 91, 92
scale, 88
of soil, 91, 92
of solutions, 89, 90
phosphate (fertilizer), 94
phosphoric acid, 89
phosphorus, 49, 69
phosphorus pentoxide, 88–90
photosynthesis, 73
plants, 7, 25
plastics, 7, 9, 55
platinum, 41, 88
pollution (by smoke), 68
polythene, 7, 55
polyvinyl chloride (PVC), 55
potassium, 58, 74, 100, 106
potassium hydroxide, 100
potassium iodide, 56, 58, 59
potassium oxide, 100
potassium permanganate, 37, 85
Priestley, 71
products, 31, 41
of electrolysis, 59
properties, 9, 10, 15, 23, 26, 43, 45
proteins, 24, 49, 74, 86, 104
Proust, 46
prussic acid, 90
purity, 11, 15, 26, 47

quicklime (see also *calcium oxide*), 91, 94

rate of reaction, 36
effect of temperature, 101
raw materials, 7, 8
Rayon, 94
react, 44
reaction (see also *chemical reaction*), 46
nuclear, 70
red lead, 33–35, 85
redox, 102, 103
reduction, 103
refine, 7
residue, 16, 33
respiration (see also *breathing*), 73, 74, 102
reversible reaction, 33
rockets (see *space rockets*)
rocks, 8, 48
rock salt, 23
rusting, 76
cause, 75
cost, 76
of metals, 74
prevention, 76

rate of, 36
and redox, 102

saliva, 92, 93
salt (see *sodium chloride*)
salt water, 15, 17, 19
sand, 16, 34, 42, 47, 48
sandstone, 49
saturated, 19
sea, 8, 11, 15, 20, 26
sea water (see *sea*)
sediment, 15, 17
selenium, 44
semi-conductors, 48, 55
separation, 15–26
by centrifuge, 17, 18
by chromatography, 23–25
by crystallization, 19
by distillation, 19–22
by evaporation, 19, 20
by filtration, 16–18
by floatation, 22, 23
separating funnel, 22
silicon, 48, 55
silicon oxide (see also *sand*), 42
silver, 7–9, 42, 47, 55, 106
in electroplating, 60
uses, 61
slaked lime (see also *calcium hydroxide*), 94
smell, 10
smoke, 11, 15, 16, 33
in explosions, 67
pollution, 68
smokeless fuels, 68
smokeless zones, 68
soap, 7, 11, 60
manufacture, 94, 104
soda lime, 66
sodium, 8, 36, 74
manufacture, 59, 106
reaction with oxygen, 88
reaction with water, 99, 100
sodium carbonate, 34, 60
sodium chlorate, 60
sodium chloride, 8, 11, 17–19, 23, 34, 56, 59, 60
sodium hydroxide, 60, 94, 100
sodium hypochlorite, 60
sodium oxide, 36, 88, 100, 105
soil, 47, 48, 92
solid, 10, 32, 42
soluble, 18, 19, 23, 25
solute, 18, 19, 23
solution, 17–20
solvent, 18–20, 23
soot (see also *smoke*), 11
sour, 90, 92
space rockets, 67, 82, 83, 104
spectroscope, 103
spectrum, 103
standard atmospheric pressure, 26
state, 10, 32
states of matter, 10, 11
state symbols, 32
steam
bath, 18, 19
in manufacture of hydrogen, 104
reaction with metals, 101, 102
steel, 7, 48, 74, 76, 94
sublimation, 31, 32
substance, 7, 9–11, 15, 31, 34
sugar, 8, 17, 19, 35, 41, 44, 56, 57, 92, 93
sulphides, 88, 105
sulphur, 31–34, 42, 45, 46, 86
sulphur dioxide, 86
sulphuric acid, 86, 90
electrolysis, 41, 57, 58
in preparation of hydrogen, 104
uses, 94
sulphurous acid, 89
superphosphate, 94
suspension, 15–18
synthesis, 33, 34, 36

taste, 18, 90
tap water, 20
temperature (effect on reaction rate), 101
temporary change, 31, 33
terminals, 41, 44, 56
Terylene, 7
thermal decomposition, 33
thermit reaction, 102
test for
acids and alkalis, 88, 89
carbon dioxide, 34
hydrogen, 42, 67
oxygen, 35
water, 33
tin, 60, 94
titanium, 102
TNT, 94
toast, 33
tooth decay, 92, 93
transistors, 48, 55

universal indicator, 88, 89, 91, 99
Universe, 8, 103
urine, 25

vapour, 11, 19, 20, 32
vegetable
dyes as indicators, 90
pH for growth of, 92
Venn diagram, 89
vinegar (see also *acetic acid*), 57, 90, 91
washing soda (see also *sodium carbonate*), 60
water, 11, 33–35, 41, 46, 48
% in air, 73
electrolysis, 41, 56
reaction with metals, 99–101
in rusting, 75, 76
test for, 33
water gas, 104
water of crystallization, 76
water vapour, 19, 20
weather balloons, 72
weedkillers, 60
weight (change in), 34
welding
oxy-acetylene, 83
thermit, 102
whisky, 21, 22
wine, 21, 72, 90
wool, 7

xenon, 72

yeast, 17

zinc, 56
cleaning of, 94
in galvanizing, 76, 105
reaction with oxygen, 88
zinc blende, 105
zinc oxide, 33